sona
BOOKS

WELCOME

Journey back to the age of dinosaurs and uncover the secrets of some of the prehistoric world's most remarkable beasts. From the Tyrannosaurus rex and Diplodocus to the Triceratops and Stegosaurus, get up close and discover how these fascinating creatures lived, hunted, evolved and ultimately died out.

In *Discovering Dinosaurs* we've gathered together some of the most fascinating articles to bring you everything you need to know about these incredible creatures that roamed our Earth millions of years ago. Why did Stegosaurus travel in herds? Could the dinosaurs have survived the asteroid that wiped them out? Is it possible to clone a dinosaur? Turn the page to find the answers to these questions and many more.

ENJOY!

CONTENTS

44

52

34

98

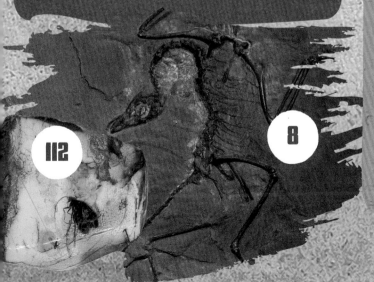

A BRIEF HISTORY OF DINOSAURS

What creatures roamed the land, sea, and air of prehistoric Earth?

Most people think of dinosaurs as big, ferocious, and extinct reptiles. That's largely true, but there are some misconceptions. Dinosaurs came in all shapes and sizes. Dinosaurs were the largest land animals of all time, but a great number of dinosaurs were smaller than a turkey.

Dinosaurs first appeared between 247 and 240 million years ago. They ruled the Earth for about 175 million years until an extinction event 65.5 million years ago wiped out all of them, except for the avian dinosaurs. Scientists don't agree entirely on what happened, but the extinction likely was a double or triple whammy involving an asteroid impact, choking chemicals from erupting volcanoes, climate change, and possibly other factors. ▶

" *They ruled the Earth for about 175 million years* "

Archaeopteryx fossil discovered in Germany. This dinosaur lived during the late Jurassic Period

> " *Some of the more advanced dinosaurs had feathers or feather-like body covering* "

PTERODACTYL

FLYING ON

Only the big, classic dinosaurs are extinct. Most experts believe that birds are living dinosaurs. Think of that next time a pigeon strafes you.

Fossils show that some of the more advanced dinosaurs had feathers or feather-like body covering, but many of them didn't fly, and probably didn't even glide. Archaeopteryx, which was for a long time considered to be the first bird (although this status is not certain), could likely launch itself from the ground, but probably couldn't fly far. Instead, feathers likely helped these bird-like dinosaurs stay warm as juveniles, or send signals to other individuals.

Many people think extinct flying reptiles called pterosaurs were dinosaurs. They were their closest relatives, but technically not dinosaurs. Pterosaurs had hollow bones, relatively large brains and eyes, and of course, the flaps of skin extending along their arms, which were attached to the digits on their front hands. The family includes Pterodactyls, with elaborate, bony head crests and lack of teeth. Pterosaurs survived up until the mass die-off 65 million years ago, when they went the way of the dodo, along with marine reptiles and other non-avian dinosaurs. ▶

ARCHAEOPTERYX

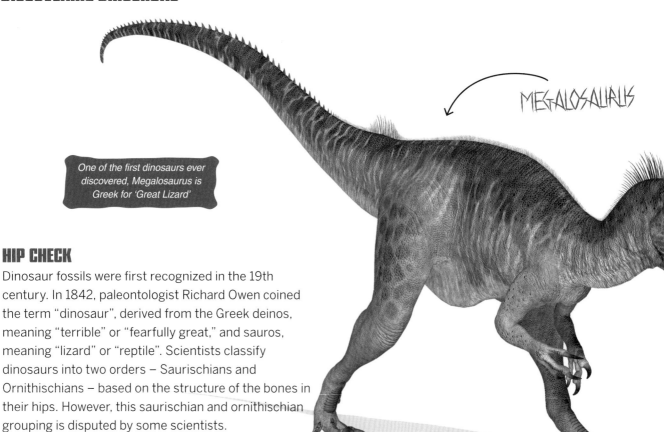

MEGALOSAURUS

> One of the first dinosaurs ever discovered, Megalosaurus is Greek for 'Great Lizard'

HIP CHECK

Dinosaur fossils were first recognized in the 19th century. In 1842, paleontologist Richard Owen coined the term "dinosaur", derived from the Greek deinos, meaning "terrible" or "fearfully great," and sauros, meaning "lizard" or "reptile". Scientists classify dinosaurs into two orders – Saurischians and Ornithischians – based on the structure of the bones in their hips. However, this saurischian and ornithischian grouping is disputed by some scientists.

Most of the well-known dinosaurs – including Tyrannosaurus rex, Deinonychus, and Velociraptor – fall into the order known as Saurischian dinosaurs. These "reptile-hipped" dinosaurs had a pelvis that pointed forward, similar to more primitive animals. They were often long-necked, had large and sharp teeth, long second fingers, and a first finger that pointed strongly away from the rest of the fingers.

Saurischians are divided into two groups – four-legged herbivores called sauropods and two-legged carnivores called

theropods (living birds are in the theropod lineage).

Theropods walked on two legs and were carnivorous. "Theropod" means "beast-footed", and they are some of the most fearsome and recognizable dinosaurs – including Allosaurus and T. rex.

Scientists have wondered whether large theropods – such as Giganotosaurus and Spinosaurus – actively hunted their prey, or simply scavenged carcasses. The evidence points to the animals working together as opportunistic hunters: they would bring down prey, but also eat dead animals that were lying around. When fossil-hunters found bones with bite marks on them, they wondered if theropods engaged in cannibalism. It appears now that the animals may have scavenged their own kind, but they didn't hunt down their own.

Sauropods were herbivores with long heads, long necks, and long tails. They were among the largest land animals ever, but they likely had small brains. The

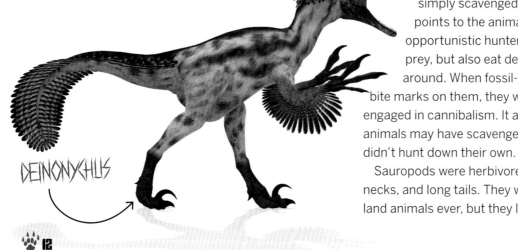

DEINONYCHUS

gentle giants like leaf-eating Apatosaurus, Brachiosaurus, and Diplodocus are part of this family.

ORNITHISCHIA

Ornithischian dinosaurs, a group that includes horned and frilled Triceratops, spiked Stegosaurus, and armored Ankylosaurus, were more mild-mannered plant eaters.

These dinosaurs were beaked herbivores. Smaller than the sauropods, the ornithischia (meaning "bird-hipped") often lived in herds, and were prey to the larger species of dinosaurs. Interestingly, the ornithischia shifted from a two-legged to a four-legged posture at least three times in their evolutionary history, and scientists think they could adopt both postures early in their evolutionary history. ▶

ANKYLOSAURUS

AEGYPTOSAURUS

Sauropods were among the largest animals to have ever walked the Earth

DISCOVERING DINOSAURS

FAMILY TREE UPDATE

In 2017, a metaphorical bombshell hit the paleontology world regarding the dinosaur family tree. A study published in the journal Nature suggested that this hip-oriented classification was incorrect. Rather, theropods are likely close cousins with the ornithischian dinosaurs, and the two groups – the theropods and Ornithischia – form a newly identified group known as Ornithoscelida, the researchers said.

The finding came about after researchers realized that theropods and Ornithischia had many anatomical features in common. If the updated tree is correct, it may explain why both theropods and Ornithischia had feathers, while other dinosaurs didn't.

However, this hypothesis will need to be tested and retested over the next several years before the paleontology community can fully accept it.

MARINE REPTILES

During the dinosaur age, a lot was happening below the oceans' surface. The "fish flippers," or ichthyopterygia, includes Ichthyosaurus — the streamlined, tuna- and dolphin-shaped ocean-going predators. This family of marine reptiles largely went extinct at the end of the Jurassic period.

OLD TREE

ORNITHISCHIA
THEROPODA
SAUROPODOMORPHA
SAURISCHIA
DINOSAURIA

NEW TREE

SAURISCHIA (PARAPHYLETIC)

SAUROPODOMORPHA
ORNITHISCHIA
THEROPODA
ORNITHOSCELIDA
DINOSAURIA

Credit: University of Cambridge

> 66 *It would be incredibly difficult, if not impossible, to clone a dinosaur* 99

BRACHIOSAURUS

Brachiosaurus comes from the Greek for 'arm lizard', as its forelegs were longer than the hind

KHTHYOSAURUS

LIOPLEURODON

The Liopleurodon's needle-like teeth and powerful jaw meant few could escape its bite

DINOSAUR CLONING

Despite the popularity of the Jurassic Park franchise, it would be incredibly difficult, if not impossible, to clone a dinosaur. In order to do this, researchers would need dinosaur DNA. But there is no known surviving DNA on record (the oldest recovered and authenticated DNA sample belongs to a 700,000-year-old horse that lived in ancient Canada).

However, some organic dinosaur matter has withstood the test of time. Researchers have uncovered a number of soft tissues from the Mesozoic era, including 80-million-year-old blood vessel belonging to a duck-billed dinosaur and 130-million-year-old proteins in an early bird fossil. But blood vessels and proteins, unlike DNA, cannot be used to clone animals.

Dinosaur eggs were incubated by heat from rotting vegetation placed in the nest

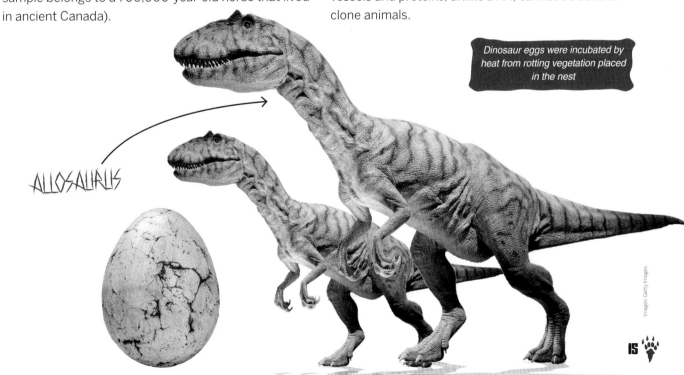

ALLOSAURUS

7 SURPRISING DINOSAUR FACTS

Triceratops ate plants. T.rex had short arms. Brontosaurus was really apatosaurus. That may be enough dinosaur knowledge to get you through a round of *Who Wants To Be a Millionaire?* or *Trivial Pursuit*, but in recent years, paleontologists have turned up many more surprising dinosaur facts.

We spoke with Mark A. Norell, the American Museum of Natural History's chairman and curator of the museum's division of paleontology, about the most surprising and little-known dinosaur facts. Here's a sneak peek: baby dinosaurs were really cute.

PIGEON-SIZED DINOSAUR

Although the museum's new exhibit focuses on large dinosaurs, not all species were massive; some were actually pretty small. In fact, "many were cat or even pigeon-size," Norell says. The smallest known pterodactyl, the Nemicolopterus crypticus (discovered in 2008), had a wingspan of only ten inches.

BABY FACE

Like most baby animals, dinosaur tots were baby-faced. In 2010, researchers found the skull* of a juvenile plant-eating dinosaur, a discovery that suggested some young dinosaurs had proportionally larger eyes and smaller faces than their parents.

* Left, not actual skull

FLUFFY DINOS

While most illustrations of dinosaurs depict them as having scaly or thick leathery skin similar to that of modern-day elephants, it was actually common for dinosaurs to have protofeathers, or a feather-like covering. Protofeathers weren't necessarily a marker for flight, however, as flightless dinos, including the velociraptor and beipiaosaurus, had downy fluff but no wings.

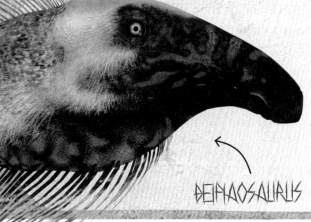

DEINAOSAURUS

4 GROWTH SPURTS

Dinosaurs grew relatively quickly. For example, the mamenchisaurus, a four-legged, plant-eating dinosaur with a long neck that made up half of its total 70-foot body length, took only about 30 years to grow to adult size.

MAMENCHISAURUS

5 LONG IN THE TOOTH

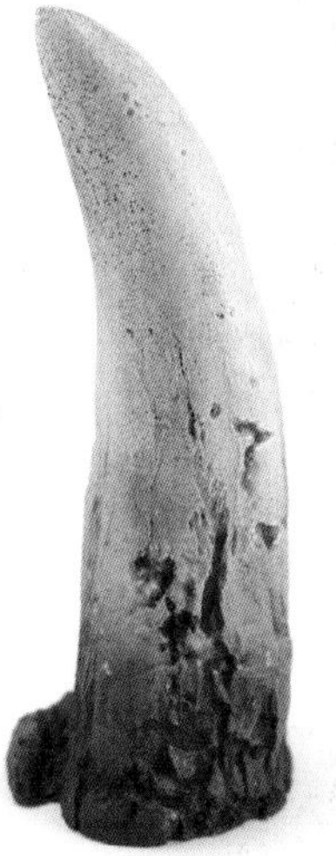

At approximately 45 feet long and 14,000 pounds, the T. rex is one of the largest land carnivores of all time, and must have looked pretty imposing. This massive monster also boasted the longest teeth. Including the root, a T. rex's tooth could be nearly ten inches long, or the length of an iPad. With 50 to 60 of those enormous teeth set in its four-foot-long jaw, the T. rex could bite off 500 pounds in a single chomp — about the weight of an adult male tiger.

6 PIECING IT TOGETHER

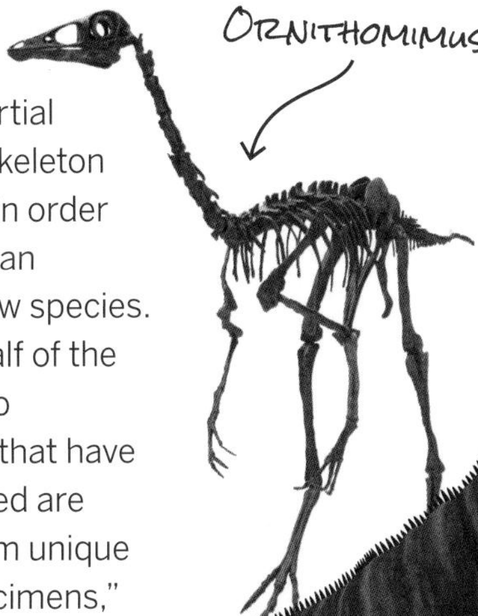

Only one complete or even partial dinosaur skeleton is needed in order to identify an entirely new species. "Almost half of the 1,200 or so dinosaurs that have been named are known from unique single specimens," Norell says.

ORNITHOMIMUS

7 THEY ROAM AMONG US

"Dinosaurs are not extinct; we just call them birds," Norell says. "In fact, birds are more closely related to dinosaurs like the T.rex than the T.rex is to sauropods like the mamenchisaurus."

EVOLUTION

Discover the story of the ever-changing world of dinosaurs

PTEROSAUR

PLATEOSAURUS

"Life outside of the oceans began to diversify during the Triassic period"

SILESAURUS

HERRERASAURUS

34

20

TRIASSIC 252–201 MILLION YEARS AGO   JURASSIC 201–145 MILLION YEARS AGO

"*During the Mesozoic era giant reptiles, dinosaurs and other monstrous beasts roamed Earth*"

TYRANNOSAURUS REX   MONOLOPHOSAURUS

CRETACEOUS 145–66 MILLION YEARS AGO

MESOZOIC ERA: AGE OF THE DINOSAURS

Life on Earth diversified rapidly during the Mesozoic era

During the Mesozoic, or "Middle Life" era, life diversified rapidly, and giant reptiles, dinosaurs and other monstrous beasts roamed Earth. The period, which spans from about 252 million to about 66 million years ago, was also known as the Age of Reptiles, or the Age of Dinosaurs.

English geologist John Phillips, the first person to create the global geologic timescale, first coined the term "Mesozoic" in the 1800s. Phillips found ways to correlate sediments found around the world to specific time periods, said Paul Olsen, a geoscientist at the Lamont-Doherty Earth Observatory at Columbia University in New York.

The Permian-Triassic boundary, at the start of the Mesozoic, is defined relative to a particular section of sediment in Meishan, China, where a type of extinct, eel-like creature known as a conodont first appeared, according to the International Commission on Stratigraphy.

The end boundary for the Mesozoic era, the Cretaceous-Paleogene boundary, is defined by a 20-inch (50 centimeters) thick sliver of rock in El Kef, Tunisia, which contains well-preserved fossils and traces of iridium and other elements from the asteroid impact that wiped out the dinosaurs. The Mesozoic era is divided up into the Triassic, Jurassic, and Cretaceous periods.

Fossils not only teach us what dinosaurs looked like, but also how they moved and what they ate

THE TRIASSIC PERIOD

The Triassic period began following the great Permian Extinction event

The Triassic period was the first phase of the Mesozoic era, and occurred around 251 million and 199 million years ago. It followed the great mass extinction at the end of the Permian period, and was a time when life outside of the oceans began to diversify.

At the beginning of the Triassic, most of the continents were concentrated in the giant C-shaped supercontinent known as Pangaea. The climate was generally very dry over much of Pangaea, with very hot summers and cold winters in the continental interior. A highly seasonal monsoon climate prevailed nearer to the coastal regions. Although the climate was more moderate farther from the equator, it was warmer than today, with no polar ice caps. Late in the Triassic, seafloor spreading in the Tethys Sea led to rifting between the northern and southern portions of Pangaea, which began the separation of Pangaea into two continents, Laurasia and Gondwana, which would be completed in the Jurassic period. ▶

TANYSTROPHEUS

Tanystropheus would hunt in shallow water as well as search for prey on land

" Most of the continents were concentrated in the giant C-shaped supercontinent known as Pangaea "

Images: Getty Images, Alamy (main)

Permian Period
225 million years ago

Pangea

Triassic Period
200 million years ago

Laurasia

Gondwana

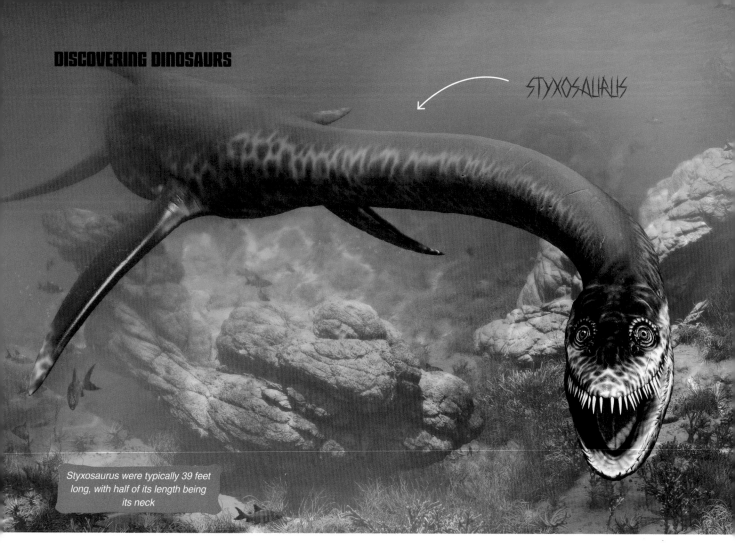

STYXOSAURUS

Styxosaurus were typically 39 feet long, with half of its length being its neck

MARINE LIFE

The oceans had been massively depopulated by the Permian Extinction, when as many as 95 percent of extant marine genera were wiped out by high carbon dioxide levels. Fossil fish from the Triassic period are very uniform, which indicates that few families survived the extinction. The mid to late Triassic period shows the first development of modern stony corals and a time of modest reef-building activity in the shallower waters of the Tethys near the coasts of Pangaea.

Early in the Triassic, a group of reptiles, the order Ichthyosauria, returned to the ocean. Fossils of early ichthyosaurs are lizard-like and clearly show their tetrapod ancestry. Their vertebrae indicate that they probably swam by moving their entire bodies side to side, like modern eels. Later in the Triassic, ichthyosaurs evolved into purely marine forms with dolphin-shaped bodies and long-toothed snouts. Their vertebrae indicate that they swam more like fish, using their tails for

propulsion with strong fin-shaped forelimbs and vestigial hind limbs. These streamlined predators were air breathers, and gave birth to live young. By the mid-Triassic, the ichthyosaurs were dominant in the oceans. One genus, Shonisaurus, measured more than 50 feet long (15 meters) and probably weighed close to 30 tons (27 metric tons). Plesiosaurs were also present, but not as large as those of the Jurassic period.

OPHTHALMOSAURUS

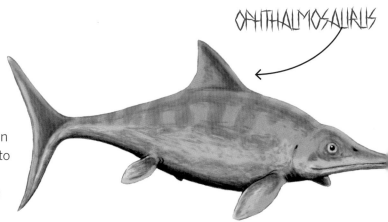

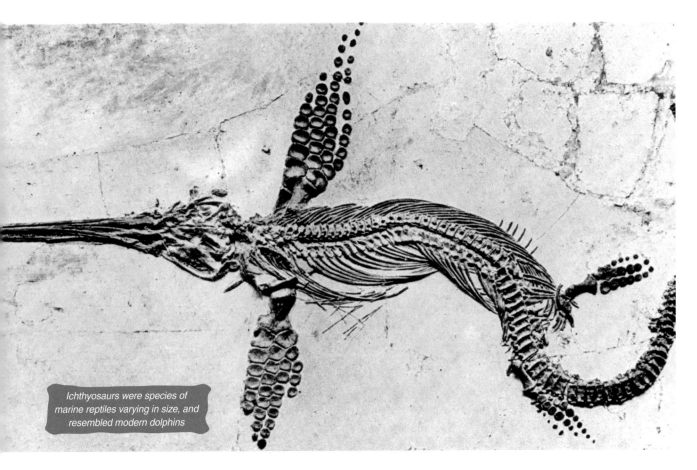

Ichthyosaurs were species of marine reptiles varying in size, and resembled modern dolphins

PLANTS AND INSECTS

Plants and insects did not go through any extensive evolutionary advances during the Triassic. Due to the dry climate, the interior of Pangaea was mostly desert. In higher latitudes, gymnosperms survived and conifer forests began to recover from the Permian Extinction. Mosses and ferns survived in coastal regions. Spiders, scorpions, millipedes, and centipedes endured, as did the newer groups of beetles. The only new insect group in the Triassic was the grasshopper.

REPTILES

The Mesozoic era is often known as the Age of Reptiles. Two groups of animals survived the Permian Extinction: Therapsids, which were mammal-like reptiles, and the more reptilian Archosaurs. In the early Triassic, it appeared that the Therapsids would dominate the new era. One genus, Lystrosaurus, has been called the Permian/Triassic "Noah", as fossils of ▶

> *Due to the dry climate, the interior of Pangaea was mostly desert*

GRASSHOPPER

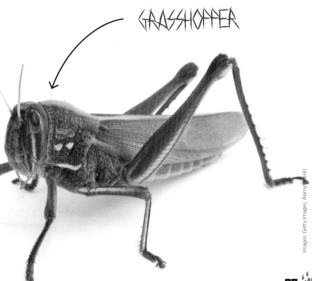

Images: Getty Images, Alamy (fossil)

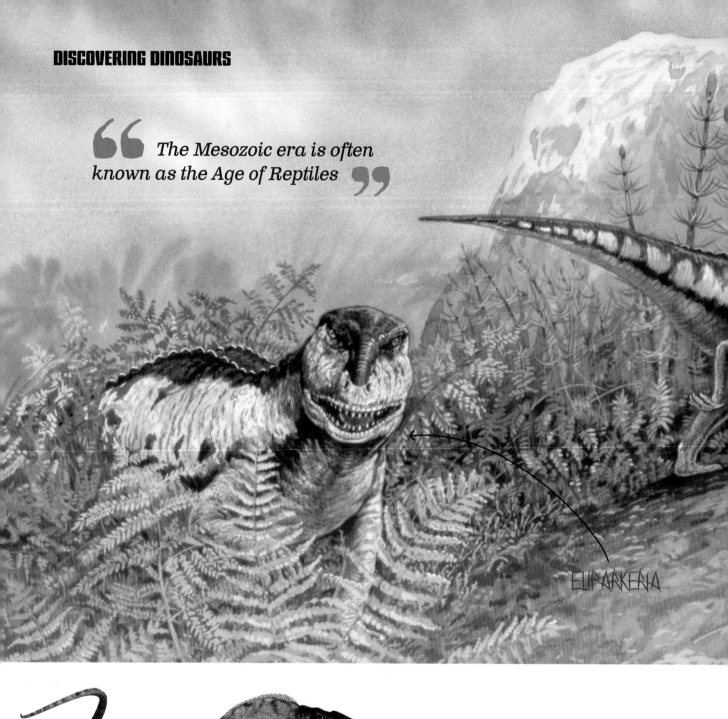

> *The Mesozoic era is often known as the Age of Reptiles*

EUPARKERIA

COELOPHYSIS

PRESTOSUCHUS

EACIOLACEATA

The Euparkeria and Ericiolacerta
lived during the Triassic period in
southern Africa

upright stance but are differentiated from true dinosaurs by the way that the pelvis and femur were arranged.

Another lineage of Archosaurs evolved into true dinosaurs by the mid-Triassic. One genus, Coelophysis, was bipedal. Although smaller than the Rauisuchians, they were probably faster, as they had a more flexibly jointed hip. Coelophysis also picked up speed by having lightweight hollow bones. They had long, sinuous necks, sharp teeth, clawed hands and a long, bony tail. Coelophysis fossils found in large numbers in New Mexico indicate that the animal hunted in packs. Some of the individuals found had remains of smaller members of the species inside the larger animals. Scientists are unclear as to whether this indicates internal gestation or possibly cannibalistic behavior.

By the late Triassic, a third group of Archosaurs had branched into the first pterosaurs. Sharovipteryx was a glider about the size of a modern crow with wing membranes attached to long hind legs. It was obviously bipedal, with tiny, clawed front limbs that were probably used to grasp prey as it jumped and glided from tree to tree. Another flying reptile, Icarosaurus, was much smaller, only the size of a hummingbird, with wing membranes sprouting from modified ribs.

EARLIEST MAMMALS

The first mammals evolved near the end of the Triassic period from the nearly extinct Therapsids. Scientists have some difficulty in distinguishing where exactly the dividing line between Therapsids and early mammals should be drawn. Early mammals of the late Triassic and early Jurassic were very small, rarely more than a few inches in length. They were mainly herbivores or insectivores, and therefore were not in direct competition with the Archosaurs or later dinosaurs. Many of them were probably at least partially arboreal and nocturnal as well. Most, such as the shrew-like Eozostrodon, laid eggs, although they clearly had fur and suckled their young. They had three ear bones like modern mammals, and a jaw with both mammalian and reptilian characteristics.

this animal predate the mass extinction, but are also commonly found in early Triassic strata. However, by the mid-Triassic, most of the Therapsids had become extinct, and the more reptilian Archosaurs were clearly dominant.

Archosaurs had two temporal openings in the skull and teeth that were more firmly set in the jaw than those of their Therapsid contemporaries. The terrestrial apex predators of the Triassic were the Rauisuchians, an extinct group of Archosaurs. In 2010, the fossilized skeleton of a newly discovered species, Prestosuchus chiniquensis, measured more than 20 feet (six meters) in length. Unlike their close relatives, the crocodilians, Rauisuchians had an

Images: Alamy, Getty Images (main)

With the Earth cooling, plant life grew, meaning herbivores got bigger, but so did their predators

THE JURASSIC PERIOD

As the oceans grew larger, so did the dinosaurs within, but the land had its fair share of giants too

The Jurassic period was the second segment of the Mesozoic era. It occurred between 199.6 and 145.5 million years ago, following the Triassic period and preceding the Cretaceous period.

During the Jurassic period, the supercontinent Pangaea split apart. The northern half, known as Laurentia, was splitting into landmasses that would eventually form North America and Eurasia,

opening basins for the central Atlantic and the Gulf of Mexico. The southern half, Gondwana, was drifting into an eastern segment that would form Antarctica, Madagascar, India, and Australia, and a western portion that would form Africa and South America. This rifting, along with generally warmer global temperatures, allowed for diversification and dominance of the reptiles known as dinosaurs. ▶

ZUPAYSAURUS ROUGIER

Jurassic Period
150 million years ago

PLANT LIFE

By the Mesozoic era, living things had evolved the capability of living on the land rather than being confined to the oceans. By the beginning of the Jurassic, plant life had evolved from Bryophytes, the low-growing mosses and liverworts that lacked vascular tissue and were confined to swampy moist areas.

Ferns and gingkoes, complete with roots and vascular tissue to move water and nutrients and a spore system of reproduction, were the dominant plants of the early Jurassic. During the Jurassic, a new method of plant reproduction evolved. Gymnosperms, cone-bearing plants such as conifers, allowed for wind distribution of pollen. This bisexual reproduction allowed for greater genetic combination and by the end of the Jurassic, the gymnosperms were widespread. True flowering plants did not evolve until the Cretaceous.

AGE OF THE DINOSAURS

As Steven Spielberg's 1993 film *Jurassic Park* asserts, reptiles were the dominant animal life forms during the Jurassic period. Reptiles had overcome the evolutionary hurdles of support and reproduction that limited the amphibians. Reptiles had strong ossified skeletons supported by advanced muscular systems for body support and locomotion. Some of the largest animals ever to live were dinosaurs of the Jurassic period. Reptiles were also capable of laying amniotic eggs, which kept the developing young moist and nourished during gestation. This allowed for the first fully terrestrial animal life cycles.

Sauropods, the "lizard-hipped" dinosaurs, were herbivorous quadrupeds with long necks balanced by heavy tails. Many, such as Brachiosaurus, were huge.

BRACHIOSAURUS

Jurassic Park, released in 1993 birthed a new generation of dinosaur fans

Some genera obtained lengths greater than 100 feet and weighed over 100 tons, making them the largest land animals ever to walk the earth. Their skulls were relatively small, with nostrils carried high near their eyes. Such small skulls meant that they had very small brains as well. Despite their small brains, this group was very successful during the Jurassic period, and had a wide geographic distribution. Sauropod fossils have been found on every continent except Antarctica. Other well-known dinosaurs of the Jurassic include the plated Stegosaurus and the flying Pterosaurs.

Carnosaurus means "meat-eating dinosaur." With such large herbivorous prey animals, it makes sense that large predators were also common. Allosaurus was one of the most populous Carnosaurs in North America; numerous intact skeletons have been found in the fossil beds of Utah. Allosaurus was superficially similar to the later evolving Tyrannosaurus rex, although cladistic analysis shows them to be only distantly related. Allosaurus was a bit smaller, with a longer jaw and heavier forelimbs. They relied on the stronger hind limbs for a running gait, but it is unclear how fast they could move.

DIMORPHODON

66 *Some of the largest animals ever to live were dinosaurs of the Jurassic period* **99**

It is unlikely to have been common for an Allosaurus to take on a healthy large adult herbivore like a Brachiosaurus or even a Stegosaurus. They were likely opportunistic, consuming young, sick, aged or ▶

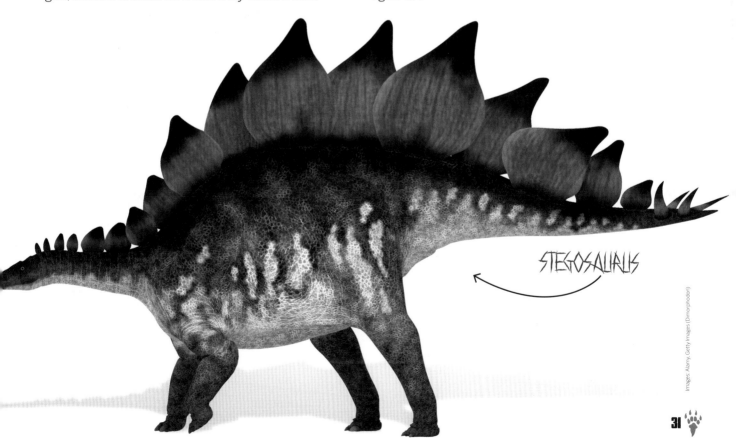

STEGOSAURUS

injured prey. They were probably able to grasp such prey with their heavily muscled forelimbs, tearing it to pieces with large claws and then swallowing the segments whole.

EARLY MAMMALS

Dinosaurs may have been the dominant land animals, but they were not alone. Early mammals were mostly very small herbivores or insectivores, and were not in competition with the larger reptiles. Adelobasileus, a shrew-like animal, had the differentiated ear and jaw bones of a mammal and dates from the late-Triassic.

In August 2011, scientists in China announced the discovery of Juramaia. This tiny animal of the mid-Jurassic has caused excitement among scientists because it is clearly a eutherian, an ancestor of placental mammals, indicating that mammals evolved much earlier than previously thought.

> *Early mammals were mostly very small herbivores or insectivores*

MARINE LIFE

Marine life of the Jurassic period was also highly diversified. The largest marine carnivores were the Plesiosaurs. These carnivorous marine reptiles typically had broad bodies and long necks with four flipper-shaped limbs. Ichthyosaurus was a more fish-shaped reptile that was most common in the early Jurassic. Because some fossils have been found with smaller individuals that appear to have been inside the larger ones, it is hypothesized that these animals may have been among the first to have internal gestation and bear live young. Cephalopod ancestors of modern squid and finned relatives of modern sharks and rays were also common. Among the most beautiful fossils of marine life were left by the spiral shells of the Ammonites.

AMMONITE FOSSIL

Allosaurus are believed to have been distantly related to the T. rex. They were opportunistic hunters

PLESIOSAURUS

Aquatic dinosaurs such as the
Plesiosaurus ruled over the
shallow waters

THE CRETACEOUS PERIOD

Higher sea levels and temperatures led to different plantlife and environments

T he Cretaceous period was the last and longest segment of the Mesozoic era. It lasted approximately 79 million years, from the minor extinction event that closed the Jurassic period about 145.5 million years ago, to the Cretaceous-Paleogene (K-Pg) extinction event dated at 65.5 million years ago.

The continents were in very different positions than they are today. Sections of Pangaea were drifting apart. The Tethys Ocean still separated the northern Laurasia continent from southern Gondwana. The North and South Atlantic were still closed, although the Central Atlantic had opened up in the late-Jurassic. By the middle of the period, ocean levels were higher; most of the landmass we are familiar with was underwater. By the end of the period, the continents were much closer to modern configuration. Africa and South America had assumed their distinctive shapes; but India had not yet collided with Asia, and Australia was still part of

" *Most of the landmass we are familiar with was underwater* "

Cretaceous Period
65 million years ago

Pterosaurs were not the only creatures in the air anymore, as ancient birds took to the skies

The number of different flowering plants grew during the Cretaceous period

CONFUCIUSORNIS

PLANT LIFE

One of the hallmarks of the Cretaceous period was the development of flowering plants. The oldest angiosperm fossil that has been found to date is Archaefructus liaoningensis, found by Ge Sun and David Dilcher in China. It seems to have been most similar to the modern black pepper plant, and is thought to be at least 122 million years old.

It used to be thought that the pollinating insects, such as bees and wasps, evolved at about the same time as the angiosperms. It was frequently cited as an example of co-evolution. New research, however, indicates that insect pollination was probably well established before the first flowers. While the oldest bee fossil was trapped in its amber prison only about 80 million years ago, evidence has been found that

WASP

bee- or wasp-like insects built hive-like nests in what is now called the Petrified Forest in Arizona.

These nests, found by Stephen Hasiotis and his team from the University of Colorado, are at least 207 million years old. It is now thought that competition for insect attention probably facilitated the relatively rapid success and diversification of the flowering plants. As diverse flower forms lured insects to pollinate them, insects adapted to differing ways of gathering nectar and moving pollen, thus setting up the intricate co-evolutionary systems we are familiar with today.

There is limited evidence that dinosaurs ate angiosperms. Two dinosaur coprolites (fossilized excrements) discovered in Utah contain fragments of angiosperm wood, according to an unpublished study presented at the 2015 Society of Vertebrate Paleontology annual meeting. This finding, as well as others, including an early Cretaceous ankylosaur that had fossilized angiosperm fruit in its gut, suggests that some paleo-beasts ate flowering plants.

Moreover, the shape of some teeth from Cretaceous animals suggests that the herbivores grazed on leaves and twigs, says Betsy Kruk, a volunteer researcher at the Field Museum of Natural History

The oldest bee fossil was trapped in its amber prison only 80 million years ago

ANKYLOSAURUS

KLIANODON

> **Theropods, including Tyrannosaurus rex, continued as apex predators**

ANIMAL LIFE

During the Cretaceous period, more ancient birds took flight, joining the pterosaurs in the air. The origin of flight is debated by many experts. In the "trees down" theory, it is thought that small reptiles may have evolved flight from gliding behaviors. In the "ground up" hypothesis, flight may have evolved from the ability of small theropods to leap high to grasp prey. Feathers probably evolved from early body coverings whose primary function, at least at first, was thermoregulation.

At any rate, it is clear that avians were highly successful, and became widely diversified during the Cretaceous. Confuciusornis (125 million to 140 million years ago) was a crow-size bird with a modern beak, but enormous claws at the tips of the wings. Iberomesornis, a contemporary – only the size of a sparrow, was capable of flight, and was probably an insectivore.

By the end of the Jurassic, some of the large sauropods, such as Apatosaurus and Diplodocus, were extinct. But other giant sauropods, including the titanosaurs, flourished, especially toward the end of the Cretaceous, Kruk says.

Large herds of herbivorous ornithischians also thrived during the Cretaceous, such as Iguanodon (a genus that includes duck-billed dinosaurs, also known as hadrosaurs), Ankylosaurus and the ceratopsians. Theropods, including T. rex, continued as apex predators until the end of the Cretaceous period. ▶

Images: Getty Images, bee in amber Co...

The asteroid created a tsunami wave, flooding much of the surrounding land

K-PG EXTINCTION EVENT

About 65.5 million years ago, nearly all large vertebrates and many tropical invertebrates became extinct in what was clearly a geological, climatic, and biological event with worldwide consequences. Geologists call it the K-Pg extinction event because it marks the boundary between the Cretaceous and Paleogene periods. The event was formally known as the Cretaceous-Tertiary (K-T) event, but the International Commission on Stratigraphy, which sets standards and boundaries for the geologic time scale, now discourages the use of the term Tertiary. The "K" is from the German word for Cretaceous: Kreide.

In 1979, a geologist who was studying rock layers between the Cretaceous and Paleogene periods spotted a thin layer of grey clay separating the two eras. Other scientists found this grey layer all over the world, and tests showed that it contained high concentrations of iridium, an element that is rare on Earth, but common in most meteorites, Kruk says.

Also within this layer are indications of "shocked quartz" and tiny glass-like globes called tektites that form when rock is suddenly vaporized then immediately cooled, as happens when an extraterrestrial object strikes the Earth with great force.

The Chicxulub crater in the Yucatan dates precisely to this time. The crater site is more than 110 miles (180 kilometers) in diameter and chemical analysis shows that the sedimentary rock of the area was melted and mixed together by

The asteroid crash caused a volcanic eruption that contributed to the mass extinctions

temperatures consistent with the blast impact of an asteroid about six miles (ten kilometers) across striking the Earth at this point.

When the asteroid collided with Earth, its impact triggered shockwaves and massive tsunamis, and sent a large cloud of hot rock and dust into the atmosphere, Kruk says. As the super-heated debris fell back to Earth, they started forest fires and increased temperatures. ▶

> *About 65.5 million years ago, nearly all large vertebrates and many tropical invertebrates became extinct*

Images: Alamy, Getty Images (main)

The asteroid crashed into a shallow sea off the coast of what is now Mexico

Within moments, everything in a 1,000km radius of the impact was in flames

VELAFRONS

"This rain of hot dust raised global temperatures for hours after the impact, and cooked alive animals that were too large to seek shelter," Kruk says. "Small animals that could shelter underground, underwater, or perhaps in caves or large tree trunks, may have been able to survive this initial heat blast."

Tiny fragments likely stayed in the atmosphere, possibly blocking part of the Sun's rays for months or years. With less sunlight, plants and the animals dependent on them would have died, Kruk says. Furthermore, the reduced sunlight would have lowered global temperatures, impairing large active animals with high-energy needs, she says.

"Smaller, omnivorous terrestrial animals, like mammals, lizards, turtles, or birds may have been able to survive as scavengers feeding on the carcasses of dead dinosaurs, fungi, roots, and decaying plant matter, while smaller animals with lower metabolisms were best able to wait the disaster out," Kruk said.

Ash covered the sky, blocking any sunlight as hot rock from the impact rained down

> "With less sunlight, plants and the animals depending on them would have died"

There is also evidence that a series of huge volcanic eruptions at the Deccan traps, located along the tectonic border between India and Asia, began just before the K-Pg event boundary. It is likely that these regional catastrophes combined to precipitate a mass extinction.

CLIMATE

The world was a warmer place during the Cretaceous period. The poles were cooler than the lower latitudes, but "overall things were warmer," Kruk told us. Fossils of tropical plants and ferns support this idea.

Animals lived all over, even in colder areas. For instance, Hadrosaurus fossils dating to the late Cretaceous were uncovered in Alaska.

When the asteroid hit, the world likely experienced so-called "nuclear winter," when particles blocked many of the Sun's rays from hitting Earth.

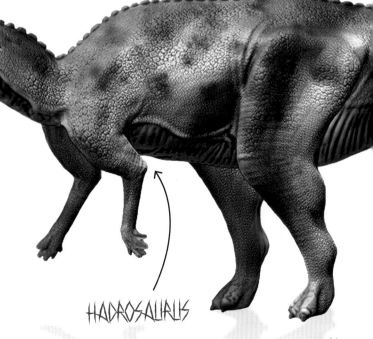

HADROSAURUS

DINOSAURS

Get up close with the prehistoric world's most fascinating beasts

44

❝ *The king of dinosaurs needed thick neck muscles to hold up its large skull and*

74

52

60

82

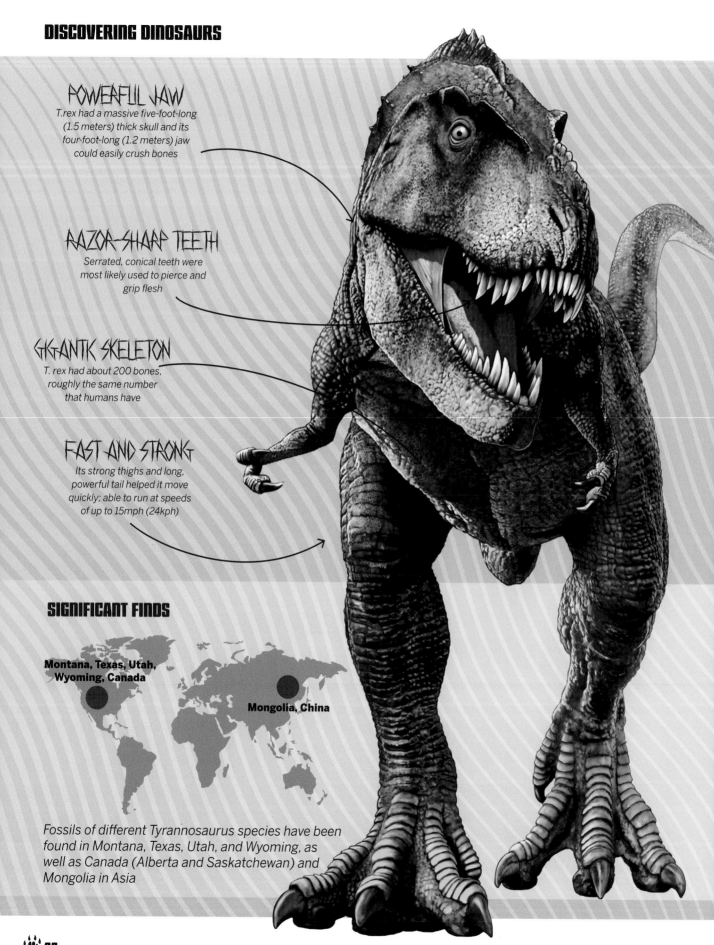

POWERFUL JAW
T.rex had a massive five-foot-long (1.5 meters) thick skull and its four-foot-long (1.2 meters) jaw could easily crush bones

RAZOR-SHARP TEETH
Serrated, conical teeth were most likely used to pierce and grip flesh

GIGANTIC SKELETON
T. rex had about 200 bones, roughly the same number that humans have

FAST AND STRONG
Its strong thighs and long, powerful tail helped it move quickly; able to run at speeds of up to 15mph (24kph)

SIGNIFICANT FINDS

Montana, Texas, Utah, Wyoming, Canada

Mongolia, China

Fossils of different Tyrannosaurus species have been found in Montana, Texas, Utah, and Wyoming, as well as Canada (Alberta and Saskatchewan) and Mongolia in Asia

TYRANNOSAURUS REX

This carnivorous predator lived during the upper Cretaceous period, 67 to 65 million years ago, toward the end of the Mesozoic era

Aside from being one of the largest of the known carnivorous dinosaurs, Tyrannosaurus rex – T. rex for short – is the dinosaur that has arguably received the most media exposure. It had a starring role in the *Jurassic Park* movies, and has a renowned exhibit at the American Museum of Natural History in New York.

The name Tyrannosaurus rex means "king of the tyrant lizards": "tyranno" means tyrant in Greek; "saurus" denotes lizard in Greek, and "rex" means "king" in Latin. In 1905, Henry Fairfield Osborn, president of the American Museum of Natural History at the time, christened it the Tyrannosaurus rex.

T. rex was a member of the Tyrannosauroidea family of huge predatory dinosaurs with small arms and two-fingered hands. Aside from Tyrannosaurs, other Tyrannosaurid genera include Albertosaurus, Alectrosaurus, Alioramus, Chingkankousaurus, Daspletosaurus, Eotyrannus, Gorgosaurus, Nanotyrannus (a controversial genus that might, in fact, be an adolescent T. rex), Prodeinodon, Tarbosaurus and Zhuchengtyrannus.

T. rex fossils are found in western North America, from Alberta to Texas. But it's possible that T. rex was an invasive species from Asia, according to a 2016 study published in Scientific Reports. An analysis of T. rex's skeletal features showed that the dinosaur king was more similar to two Tyrannosaurs in Asia, Tarbosaurus and Zhuchengtyrannus, than it was to North American Tyrannosaurs, the researchers told us. They also believe that the paleo beast crossed over about 67 million years ago when the seaway between Asia and North America receded. ▶

T. REX FACTS AND STATS

Length	About 40 feet (12 meters)
Height	15 to 20 feet tall (4.6 to 6 meters)
Weight	Up to nine tons (about 8,164kg)
Diet	Meat; primarily ate herbivorous dinosaurs, including the Edmontosaurus and the Triceratops

SIZE COMPARISON

PERIOD		Triassic		Jurassic		Cretaceous	Age of Mammals	
Millions of years ago	250		200		145		65	Present

The largest T. Rex skeleton discovered, Sue, stood at 40 feet long and 13 feet tall

However, they added that the finding is still preliminary, and other experts maintain that T. rex evolved in North America.

AN IMPOSING FIGURE

The largest and most complete T. rex skeleton ever found was nicknamed Sue, after the woman who discovered it, paleontologist Sue Hendrickson. Measurements suggest that T. rex was one of the largest carnivorous dinosaurs ever, coming in at up to 13 feet (four meters) tall at the hips (its highest point, since it did not stand erect) and 40 feet (12.3 meters) long. A recent analysis of Sue, published in the journal PLOS ONE in 2011, shows that T. rex weighed as much as nine tons (8,160 kilograms).

T. rex had strong thighs and a powerful tail, which counterbalanced its large head (Sue's skull is five feet, or 1.5 meters, long) and allowed it to move quickly. The 2011 study, which also modeled T. rex's muscle distribution and center of mass, suggests that the giant could run 10 to 25 mph (17 to 40 km/h), as previous studies had estimated.

Its two-fingered forearms were puny, making it unlikely that T. rex could use them to kill or even get a meal to its mouth. However, it's possible that T. rex had such tiny arms because of its powerful bite, according to research from Michael Habib, an assistant professor of clinical cell and neurobiology at the University of Southern California and a research associate at the Dinosaur Institute at the Natural History Museum of Los Angeles County.

The king of dinosaurs needed thick neck muscles to hold up its large skull and power its forceful bite. Neck and arm muscles compete for space in the shoulder, and it appears that the neck muscles edged out the arm muscles in T. rex's case, according to

The T. rex's arm had two large claws that were used to cut deep into its prey

> *Having short arms may have been beneficial to the king in the long run*

Habib's research. Moreover, long arms can be broken, are vulnerable to disease, and take energy to maintain, so having short arms may have been beneficial to the king in the long run, Habib's research shows.

The real work of dispensing with its prey was left to the dinosaur's massive, thick skull. T. rex had the strongest bite of any land animal that ever lived, according to a 2012 study in the journal Biology Letters. The dinosaur's bite could exert up to 12,814 pounds-force (57,000 Newtons), which is roughly equivalent to the force of a medium-size elephant sitting down. ▶

The T. rex's arms were strong enough to grip its prey, and could lift around 200kg

" T. rex was a huge carnivore, and primarily ate herbivorous dinosaurs "

Due to its size and strength, the T. rex didn't have to compete for prey in the jungles

The mouth contained around 60 teeth, with the average size being around eight inches long

T. rex had a mouth full of serrated teeth; the largest tooth of any carnivorous dinosaur ever found was 12 inches (30 centimeters) long. But not all of the dinosaur's teeth served the same function, according to a 2012 study in the Canadian Journal of Earth Sciences. Specifically, the dinosaur's front teeth gripped and pulled; its side teeth tore flesh, and its back teeth diced chunks of meat and forced food into the throat. Importantly, the study found that T. rex's teeth were wide and somewhat dull (rather than flat and daggerlike), allowing the teeth to withstand the forces exerted by struggling prey.

T. rex may be big, but its predecessors were small. The first tyrannosaurs, which were human- to horse-size, originated about 170 million years ago during the mid-Jurassic. Though lacking in stature, these little tyrannosaurs had advanced brains and advanced sensory perceptions, including hearing, a 2016 study detailed in the Proceedings of the National Academy of Sciences journal revealed. The finding, on a newfound mid-Cretaceous tyrannosaur named Timurlengia euotica, suggests that the advanced brains tyrannosaurs developed while they were still small helped them become apex predators once they grew to T. rex's size.

WHAT DID T. REX EAT?

T. rex was a carnivore, mainly eating herbivorous dinosaurs, including Edmontosaurus and Triceratops. It acquired food through scavenging and hunting, was incredibly fast, and ate hundreds of pounds of food, says University of Kansas paleontologist David Burnham.

"T. rex was probably opportunistic, and may have fed on carcasses, but that is not a very abundant or consistent food source," Burnham told us. "T. rex had a hard life. They had to go out and kill for food when they were hungry."

For many years, the evidence that T. rex actually hunted for its meals was circumstantial, based on such things as bones with bite marks, teeth near carcasses and foot tracks suggesting pursuits, ▶

Images Getty Images. Alamy (main)

Burnham said. But in a 2013 study in the Proceedings of the National Academy of Science journal, Burnham and his colleagues unveiled direct evidence of T. rex's predatory nature: a T. rex tooth embedded in a duckbill dinosaur's tailbone, which healed over the tooth (meaning the duckbill got away).

"We found the smoking gun!" Burnham said. "With this discovery, we now know the monster in our dreams is real."

T. rex was also not above enjoying another T. rex for dinner, according to a 2010 analysis published in PLOS ONE of T. rex bones with deep gashes created by T. rex teeth. However, it's not clear if the cannibalistic dinosaurs fought to the death or merely ate the carcasses of their own kind.

Scientists are unsure whether T. rex hunted alone or in packs. In 2014, researchers found dinosaur track marks in the foothills of the Canadian Rockies in British Columbia — out of the seven tracks, three belonged to Tyrannosaurids, most likely Albertosaurus, Gorgosaurus, or Daspletosaurus. The study, published in PLOS ONE, suggests that T. rex's relatives, at least, hunted in packs.

The T. rex wasn't solely a hunter. It would scavenge off carcasses it found

66 *T. rex was not above enjoying another T. rex for dinner* 99

WHEN AND WHERE T. REX LIVED

T. rex fossils are found in a variety of rock formations dating to the Maastrichtian age of the upper Cretaceous period, which lasted from 67 to 65 million years ago, toward the end of the Mesozoic era. It was among the last of the non-avian dinosaurs to exist prior to the Cretaceous-Paleogene extinction event, which wiped out the dinosaurs.

More mobile than many other land-based dinosaurs, T. rex roamed throughout what is now western North America, at the time an island continent identified as Laramidia. More than 50 T. rex skeletons have been unearthed, according to National Geographic. Some of these remains are nearly complete skeletons, and at least one skeleton included soft tissue and proteins.

The large tail helped keep the dinosaur's balance, which was essential when chasing prey

The T. rex fossil 'Black Beauty' on display at the Royal Tyrrell Museum in Alberta, Canada

The fossil of a T. Rex's foot on display in the Museum of the Rockies in Montana, USA

Fossil hunter Barnum Brown discovered the first partial skeleton of a T. rex in the Montana portion of the Hell Creek Formation in 1902. He later sold this to the Carnegie Museum of Natural History in Pittsburgh. Another T. rex fossil discovery of his, also from Hell Creek, is on display in the American Museum of Natural History in New York.

In 2007, scientists unearthed what may be a T. rex footprint in Hell Creek, and described their discovery in the journal Palaios. If the track did indeed belong to T. rex, it would be only the second confirmed T. rex footprint ever discovered, the first being an example discovered in New Mexico in 1993.

STEGOSAURUS

These armored herbivores lived during the late Jurassic Period, about 155 million to 150 million years ago

BONY PLATES

Plates, called scutes, were made of a bony material but were not solid; they had lattice-like structures and blood vessels throughout

PROTECTIVE SPIKES

Spikes, which pointed outward from the sides of the tail, were up to four feet (1.2 meters) long and were used for protection from predators

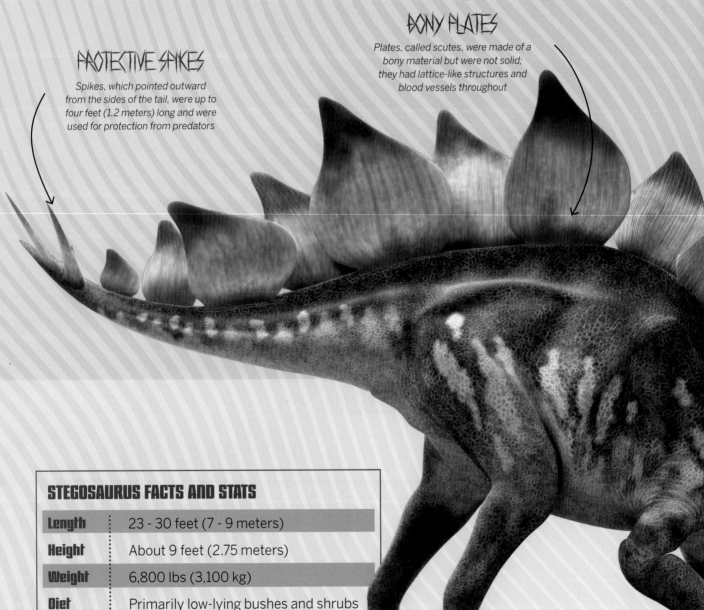

STEGOSAURUS FACTS AND STATS

Length	23 - 30 feet (7 - 9 meters)
Height	About 9 feet (2.75 meters)
Weight	6,800 lbs (3,100 kg)
Diet	Primarily low-lying bushes and shrubs and other vegetation, including ferns, mosses, cycads, fruits and conifers.

tegosaurus was a large, plant-eating dinosaur that lived during the late Jurassic Period, about 150.8 million to 155.7 million years ago, primarily in western North America. It was about the size of a bus and carried around two rows of bony plates along its back that made it appear even bigger.

Stegosaurus is a bit of a media darling because there is so much material to help scientists reconstruct its distinctive appearance. It has been depicted on television and in movies, most notably chasing Faye Wray in *King Kong* and appearing in the second and third installments of the *Jurassic Park* films. A newspaper cartoon even helped name one of its body parts.

Stegosaurus has a reputation for having a small brain and one of the lowest-brain-to-body ratios among dinosaurs. "The brain of Stegosaurus was long thought to be the size of a walnut," said armored-dinosaur expert Kenneth Carpenter, director of the USU Eastern Prehistoric Museum in Utah. "But actually, its brain had the size and shape of a bent hotdog." ▶

Fossils from about 80 individuals were discovered in the Morrison Formation, which is centered in Wyoming and Colorado.

The Stegosaurus' brain was the size and shape of a hotdog, not a walnut as originally thought

SIGNIFICANT FINDS

Western North America

Western Europe, southern India, China and southern Africa

SIZE COMPARISON

STRENGTH IN NUMBERS

Stegosaurus was the largest and most well-known of the stegosaurids, a family of armored dinosaurs. Fossils suggest they traveled in herds

PERIOD		Triassic		Jurassic		Cretaceous		Age of Mammals	
Millions of years ago	250		200		145		65		Present

Images: Getty Images

At one point scientists theorized that it had an auxiliary "brain" — not an actual brain but a bundle of nerves — above its rear legs to help control its movements because its brain was so small. This idea stemmed from the fact that the dinosaur had an enlarged canal in the pelvic region of its spinal cord, Carpenter said. However, that theory has since been rejected, and it is unclear what function this cavity served. It may have held a glycogen body, a structure found in birds that may play a role in energy storage, according to study published in the journal Paleobiology in 1990.

SIZE OF STEGOSAURUS

Stegosaurus was the largest and most well-known member of the Stegosauridae family of armored dinosaurs. The largest species was Stegosaurus armatus, a behemoth that grew up to about 30 feet (nine meters) long. It is considered a "type species," or the species that serves as the primary example of the Stegosaurus genus. However, Stegosaurus stenops — the best known and most studied Stegosaurus species due to the abundance of its

fossils, including a near-complete skeleton — may instead be more deserving of that title, Carpenter and paleontologist Peter Galton argued in respective papers in the Swiss Journal of Geosciences in 2010.

There is also considerable debate regarding how many valid species of Stegosaurus exist, Carpenter told us. On the one hand, there may be up to three or four species based on the differences seen in fossils. But if you're a "taxonomic clumper", you may think there's only a single species of Stegosaurus because a wide range of variation can exist within a single species (such as how all dog breeds belong to Canis lupus familiaris), Carpenter said.

Stegosaurus means "roofed lizard," which was derived from the belief by 19th-century paleontologists that the plates lay flat along its back like shingles on a roof. But most evidence supports the idea that the plates alternated in two rows, pointy side up from the dinosaur's neck down to its rear. Its 17 plates, called scutes, were made of a bony material called osteoderms but were not solid; they had lattice-like structures and blood vessels throughout.

Though it is uncertain what purpose the plates served, the blood vessels within the plates suggest temperature regulation (heat dissipation) was likely one function. However, that theory has been questioned — the microstructure of the plates suggests they weren't used to radiate heat, according to a 2005 analysis published in Paleobiology. A more recent study, published 2010 in the Swiss Journal of Geosciences, concluded that the plates may have played a passive role in managing body temperature because of their large size and extensive blood vessels (similar to the way a toucan's large bill naturally radiates body heat), but that wasn't their primary function.

Instead, Stegosaurus likely used its plates for display purposes. "Showing off, species recognition, attracting mates — that sort of thing," Carpenter explained. ▶

> **The plates may have played a passive role in managing body temperature**

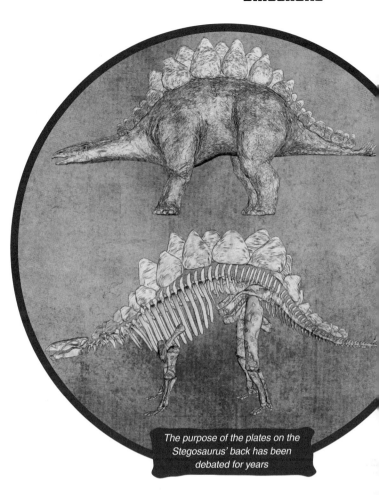

The purpose of the plates on the Stegosaurus' back has been debated for years

Stegosaurus were hunted by large theropod predators such as Allosaurus

Images: Getty Images, Alamy (skeleton)

" *Stegosaurus was the largest and most well-known member of the Stegosauridae family* **"**

Evidence from fossils indicates that the dinosaurs would travel in herds

The Stegosaurus' tail had sharp spikes and could be whipped to attack would-be predators

Despite its large size, Stegosaurus' head was relatively small and kept low to the ground

Stegosaurus also had spikes at the end of its flexible tail, which pointed outward from the sides. Scientists began informally calling the spikes thagomizers after a pop culture reference in 1982, when a *Far Side* cartoon showed a group of cavemen calling the sharp spikes thagomizers "after the late Thag Simmons," according to *New Scientist*.

Experts think these spikes were used for defense against predators because of two lines of evidence. For one thing, about ten percent of spikes found are damaged at the tip, Carpenter said. Additionally, scientists have found allosaur fossils (Stegosaurus' main predator) with puncture wounds from thagomizers.

Stegosaurus' long skull was pointed and narrow. It had an unusual head-down posture because its forelimbs were short in relation to its hind legs. ▶

Images: Getty Images, Alamy (above)

The dinosaurs were not fast travelers but they weren't defenseless creatures

This leg-length imbalance suggests that the dinosaur couldn't move very fast, because the stride of its back legs would have then overtaken its front legs.

WHAT DID STEGOSAURUS EAT?

Stegosaurus was a herbivore, as its toothless beak and small teeth were not designed to eat flesh and its jaw was not very flexible. Interestingly, unlike other herbivorous, beaked dinosaurs (including Triceratops and the duck-billed Hadrosaurids), Stegosaurus did not have strong jaws and grinding teeth. Instead, its jaws likely only allowed up and down movements, and its teeth were rounded and peg-like. It also had cheeks, which gave it room to chew and store more food than many other of dinosaurs.

As a result of its short neck and small head, it most likely ate low-lying bushes and shrubs and other vegetation, including ferns, mosses, cycads, fruits, conifers, horsetails and even fallen fruit. Some scientists believe Stegosaurus potentially could have stood on its hind legs to reach some taller plants, but this idea is debated.

In a 2010 study in the Swiss Journal of Geosciences, scientists modeled the teeth and jaws

The Stegosaurus was a herbivore, feeding off of low-hanging vegetation and fruit

The plates on the back were made from bone and likely covered in skin or toughened horn

of Stegosaurus to better understand what the dinosaur was capable of eating. The research showed that Stegosaurus had a very weak bite (weaker than a human bite) and could only break down twigs and branches less than half an inch in diameter.

FOSSIL DISCOVERIES

Fossils suggest that Stegosaurus traveled in multi-age herds.

The first Stegosaurus fossils were discovered in 1876 in Colorado by M.P. Felch and were then named by Othniel C. Marsh in 1877.

Fossils from about 80 individuals were discovered in the Morrison Formation, which is centered in Wyoming and Colorado. Stegosaurus is even the state fossil of Colorado in tribute to the number of dinosaur skeletons found in the state. The formation also reaches into Montana, North Dakota, South Dakota, Nebraska, Kansas, Oklahoma, Texas, New Mexico, Arizona, Utah and Idaho.

In Wyoming's Red Canyon Ranch in 2003, Bob Simon, president of the dinosaur excavation and preservation corporation Virginia Dinosaur Company and Dinosaur Safaris, discovered the most complete (more than 90 percent) Stegosaurus specimen to date.

And, in 2007, researchers discovered a Stegosaurus fossil in Portugal. The finding, which was published in the German science journal 'Naturwissenschaften', shows that the dinosaur lived in Europe in addition to North America, and supports the idea that the two continents were once connected by temporary land bridges that surfaced during low tide.

Images: Getty Images, Alamy (top left)

TRICERATOPS

These elephant-sized beasts roamed North America about 68 million to 65 million years ago

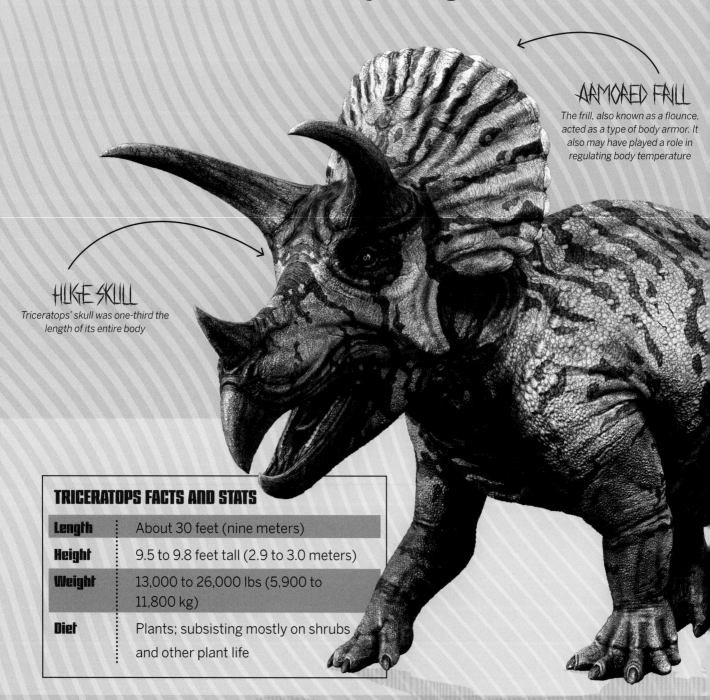

AARMORED FRILL

The frill, also known as a flounce, acted as a type of body armor. It also may have played a role in regulating body temperature

HUGE SKULL

Triceratops' skull was one-third the length of its entire body

TRICERATOPS FACTS AND STATS

Length	About 30 feet (nine meters)
Height	9.5 to 9.8 feet tall (2.9 to 3.0 meters)
Weight	13,000 to 26,000 lbs (5,900 to 11,800 kg)
Diet	Plants; subsisting mostly on shrubs and other plant life

Triceratops, with its three horns and bony frill around the back of its head, is one of the most recognizable dinosaurs. Its name is a combination of the Greek syllables tri-, meaning "three," kéras, meaning "horn," and ops, meaning "face." The dinosaur roamed its territory in North America about 68 million to 65 million years ago, during the late Cretaceous Period.

Since Triceratops' discovery in 1887, up to 16 species of the dinosaur have been proposed, but only two — T. horridus and T. prorsus — are currently considered valid, according to a 2014 study in the journal Proceedings of the National Academy of Sciences (PNAS), which found that T. horridus likely evolved into T. prorsus over a span of one million to two million years.

For the study, researchers collected and analyzed dinosaur fossils from the Hell Creek Formation in Montana, which contains lower, middle and upper geological subdivisions. The most commonly recovered dinosaur from the formation was Triceratops, said ▶

During a 2014 study, Triceratops fossils were the most commonly recovered from the Hell Creek Formation in Montana

In 1887, the first bones of a Triceratops were discovered in Denver.

SIGNIFICANT FINDS

Colorado, South Dakota, North Dakota, Wyoming, Montana and western Canada

SIZE COMPARISON

SLOW MOVERS
Triceratops could move at only ten miles per hour (16 kilometres per hour)

PERIOD	Triassic	Jurassic	Cretaceous	Age of Mammals
Millions of years ago	250	200	145	65	Present

Images: Getty Images

study first author John Scannella, a paleontologist and Triceratops expert at Montana State University.

"We started to notice that the Triceratops in the lower unit of the formation are different from those in the upper unit," Scannella told Live Science. "And Triceratops in the middle unit have a combination of features seen in individuals found in the lower and upper units." Triceratops prorsus, he said, is found in the upper unit of the formation, and specimens in the upper portion of the middle unit have more T. prorsus features and fewer T. horridus features.

There is currently significant debate about whether two other genera of Ceratopsidae (the taxonomic family Triceratops belongs to), Torosaurus and Nedoceratops (formerly Diceratops), are really distinct genera or just Triceratops specimens at different stages of life.

In a 2010 study in the Journal of Vertebrate Paleontology, Scannella and his colleague John ("Jack") Horner argued that Torosaurus, which is mainly distinguished from Triceratops by having an expanded frill with large holes in the bone, was actually Triceratops in its old age. "We found

Triceratops' brow horns stretched out in front of its snout while the nose horn was shorter

evidence that the frill on the back of the skull [of Triceratops] expands relatively late during growth," Scannella said, adding that the microstructure of Torosaurus bones suggests they are older than even the largest Triceratops specimens. "This suggested that Torosaurus, rather than being a distinct genus, was actually a fully grown Triceratops."

In a subsequent 2011 study in the journal PLOS ONE, Scannella and Horner used similar reasoning to argue Nedoceratops hatcheri, of which there is only a single specimen, is actually a transitional stage between the young Triceratops and the old Torosaurus. Again, one of the main differences between the animals in question is the frill: Torosaurus has large frill holes, which are smaller in Nedoceratops and, for the most part, absent in Triceratops (though some specimens appear to show evidence of the beginning of holes). This suggests that the holes grow over time as the frill develops and expands, they reasoned.

However, some other paleontologists contest this single-genus idea. In a 2012 PLOS ONE article, for instance, researchers presented evidence of Torosaurus bones that are not fully fused, suggesting the specimen is still immature (and, therefore, not a fully mature Triceratops). They further suggested the frill holes of Nedoceratops are pathological (related to a disease or health issue).

More fossils of Nedoceratops and a distinctly juvenile specimen of Torosaurus would settle the debate, Scannella said.

Triceratops' three horns came in handy for defending itself against predators such as the T. rex

Standing at roughly 9.5 feet tall, Triceratops were around the size of an African elephant

AN ELEPHANT-SIZE DINOSAUR

Triceratops were a massive animal, comparable in size to an African elephant, according to a 2011 article in the journal Cretaceous Research. It grew up to 30 feet (nine meters) and weighed well over 11,000 lbs (5,000 kg) — some large specimens weighed nearly 15,750 lbs (7,150 kg). ▶

> *Torosaurus has large frill holes, which are smaller in Nedoceratops and absent in Triceratops*

There has been some debate over whether Torosaurus were a distinct genus or a fully grown Triceratops

A fossil of a Triceratops' head on display at CosmoCaixa museum in Barcelona, Spain.

It is commonly believed that Triceratops were solitary creatures but this could be wrong

It had strong limbs to move and support its massive body. The forelimbs, which were shorter than the rear ones, each had three hooves; the rear limbs had four hooves each. A 2012 study in the journal Proceedings of the Royal Society B suggested that Triceratops had an upright posture like an elephant's, rather than a sprawling, elbows-out posture like a lizard's, as was initially believed when it was discovered.

The head of Triceratops was among the largest of all land animals, some making up one-third of the entire length of the dinosaur's body. The largest skull found has an estimated length of 8.2 feet (2.5 meters), according to Scannella's 2010 Journal of Vertebrate Paleontology study.

Triceratops had three horns: two massive ones above its eyes, and a smaller horn on its snout. The two brow horns appear to have twisted and lengthened as a Triceratops aged, according to a 2006 study in the journal Proceedings of the Royal Society B. During a Triceratops' juvenile years, its horns were little stubs that curved backward; as the animal continued to grow into young adulthood, the horns straightened out; finally, the horns curved forward and grew up to three feet long (one meter), probably after the dinosaur reached sexual maturity.

It is likely Triceratops' horns and frill were used in combat against other Triceratops, as well as for visual display (mating, communication and species recognition), according to a 2009 PLOS ONE study.

The dinosaur also used its horns and frill in fights against its main predator, tyrannosaurs. Paleontologists have uncovered brow horn and skull Triceratops bones that were partially healed from tyrannosaur tooth marks, suggesting the Triceratops successfully fended off its attacker, according to a study published in the book *Tyrannosaurus rex, the* ▶

" *The head of Triceratops was among the largest of all land animals* "

Images: Getty Images, Alamy (inset)

Tyrant King (2008, Indiana University Press). But T. rex bite marks on other Triceratops bones suggests the carnivore did sometimes successfully feed on the horned dinosaur, a 1996 study in the Journal of Vertebrate Paleontology suggests.

Instead of being smooth, the skin of Triceratops, at least around its tail, may have been covered in bristle-like formations, similar to the ancient ceratopsian Psittacosaurus.

WHAT DID TRICERATOPS EAT?

Triceratops was a herbivore, existing mostly on shrubs and other plant life. Its beak-like mouth was best suited for grasping and plucking rather than biting, according to a 1996 analysis in the journal Evolution. It also likely used its horns and bulk to tip over taller plants.

It had up to 800 teeth that were constantly being replenished, and were arranged in groups called batteries, with each battery having 36 to 40 tooth columns in each side of each jaw and three to five teeth per column, the Evolution study notes. It may have eaten a range of plants, including ferns, cycads and palms.

FOSSIL DISCOVERIES

In 1887, the first bones of a Triceratops were discovered in Denver and were sent to Othniel Charles Marsh. At first, Marsh believed it was a bison. It wasn't until more Triceratops bones were found in 1888 that Marsh gave the beast the name Triceratops.

> " *It is likely Triceratops' horns and frill were used in combat against other Triceratops* "

Bones have shown that some Triceratops successfully defended themselves from T. rex

Triceratops' horns and beak may have been used to bring down food from above

To date, more than 50 Triceratops skulls have been found in the Hell Creek Formation alone, according to Scannella's 2014 PNAS study.

While no complete skeleton has been unearthed, partial skeletons and skulls, including some from babies, have been found in Montana, South Dakota, North Dakota, Colorado, Wyoming and Canada (Saskatchewan and Alberta). Triceratops was confined to North America because the continent had already split from Europe and, along with South America, had begun to drift across the ocean by the time the dinosaur evolved.

Triceratops fossils have typically been discovered as solitary individuals. But in a 2009 article in the Journal of Vertebrate Paleontology, scientists reported the first discovery of a Triceratops "bonebed" that contained three juvenile remains together, suggesting a gregarious (and possibly herding) nature to the dinosaurs.

The dinosaur had plenty of teeth for grabbing and pulling vegetation with

Images: Getty Images, Alamy (tooth & fossil)

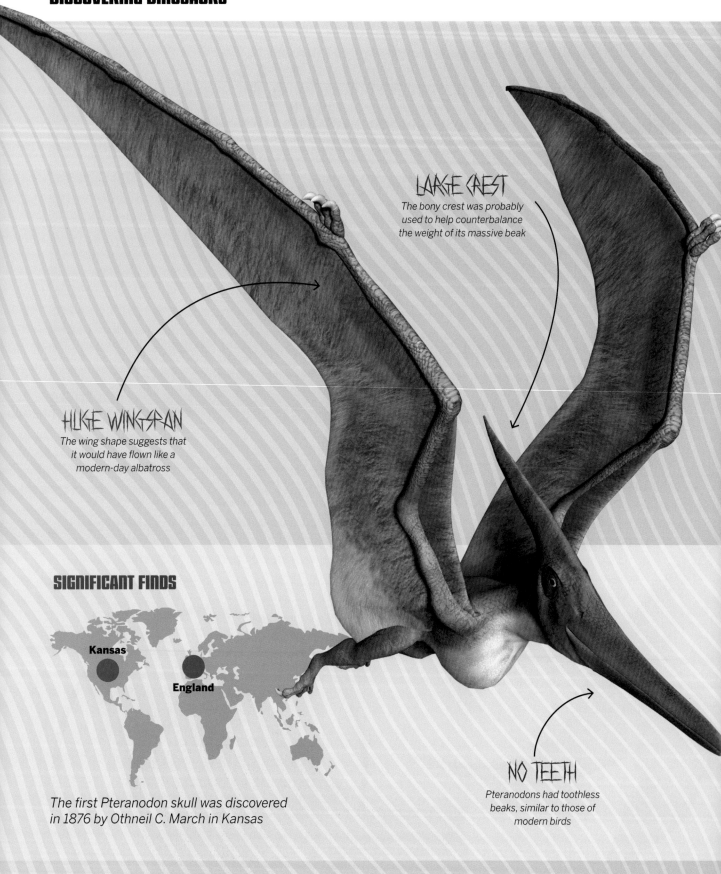

LARGE CREST
The bony crest was probably used to help counterbalance the weight of its massive beak

HUGE WINGSPAN
The wing shape suggests that it would have flown like a modern-day albatross

SIGNIFICANT FINDS

Kansas

England

The first Pteranodon skull was discovered in 1876 by Othneil C. March in Kansas

NO TEETH
Pteranodons had toothless beaks, similar to those of modern birds

PTERODACTYL, PTERANODON
& OTHER FLYING 'DINOSAURS'

THALASSODROMEUS

Pteranodon was a flying reptile that lived 75 million to 85 million years ago during the Upper Cretaceous period

Pterodactyl is the common term for the winged reptiles properly called pterosaurs, which belong to the taxonomic order Pterosauria. Scientists typically avoid using the term and concentrate instead on individual genera, such as Pterodactylus and Pteranodon.

There are at least 130 valid pterosaur genera, according to David Hone, a paleontologist at Queen Mary University of London. They were widespread and lived in numerous locations across the globe, ranging from China to Germany to the Americas.

Pterosaurs first appeared in the late Triassic period and roamed the skies until the end of the Cretaceous period (228 to 66 million years ago), according to an article published in 2008 in the German scientific journal Zitteliana. Pterosaurs lived among the dinosaurs and became extinct around the same time, but they were not dinosaurs. ▶

PTERANODON FACTS AND STATS

Length	18 feet (5.6 meters)
Height	Six feet (1.8 meters) at the hips
Weight	55 pounds (25 kilograms)
Diet	Pteranodons were primarily carnivores and fed on fish, mollusks, crabs, insects, and carcasses of dinosaurs and other animals

SIZE COMPARISON

PERIOD	Triassic	Jurassic	Cretaceous	Age of Mammals

Millions of years ago   250   200   145   65   Present

Rather, pterosaurs were flying reptiles.

Modern birds didn't descend from pterosaurs; birds' ancestors were small, feathered, terrestrial dinosaurs.

The first pterosaur discovered was Pterodactylus, identified in 1784 by Italian scientist Cosimo Collini, who thought that he had discovered a marine creature that used its wings as paddles.

A French naturalist, Georges Cuvier, proposed that the creatures could fly in 1801, and then later coined the term "Ptero-dactyle" in 1809 after the discovery of a fossil skeleton in Bavaria, Germany. This was the term used until scientists realized they were finding different genera of flying reptiles. However, "pterodactyl" stuck as the popular term.

Pterodactylus comes from the Greek word pterodaktylos, meaning "winged finger," which is an apt description of its flying apparatus. The primary component of the wings of Pterodactylus and other pterosaurs was a skin and muscle membrane stretching from their highly elongated fourth fingers to their hind limbs.

The reptiles also had membranes running between the shoulders and wrists (possibly incorporating the first three fingers of the hands), and some groups of pterosaurs had a third membrane between their legs, which may have connected to or incorporated a tail.

Early research suggested pterosaurs were cold-blooded animals that were more suited to gliding than active flying. However, scientists later

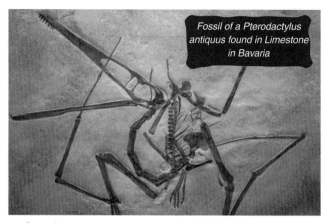

Fossil of a Pterodactylus antiquus found in Limestone in Bavaria

> **Pterodactylus comes from the Greek word pterodaktylos, meaning 'winged finger,' which is an apt description**

discovered that some pterosaurs, including Sordes pilosus and Jeholopterus ninchengensis, had furry coats consisting of hairlike filaments called pycnofibers, suggesting they were warm-blooded and generated their own body heat, according to a 2002 study in the Chinese Science Bulletin.

What's more, a 2010 study in the journal PLOS ONE suggested pterosaurs had powerful flight muscles, which they could use to walk as quadrupeds (on all fours) like vampire bats and vault into the air. Once airborne, the largest pterosaurs (Quetzalcoatlus northropi) could reach speeds of up to 80mph (130 km/h) for a few minutes and then glide at cruising speeds of about 56mph (90km/h), the study found.

SIZES OF PTEROSAURS

Pterodactylus antiquus (the only known species of the genus) was a comparatively small pterosaur, with an estimated adult wingspan of about 3.5 feet (1.06 meters), according to a 2012 study in the journal Paläontologische Zeitschrift. There was some confusion early on as to the size of the Pterodactylus, because some of the specimens turned out to be juveniles rather than adults.

Pteranodon, discovered in 1876 by Othniel C.

The Tupandactylus' crest, made from bone and soft tissue, was used to signal to other pterosaurs

Quetzalcoatlus were able to hunt both in the air and while walking on the ground

With a wingspan of ten meters, the Quetzalcoatlus was the size of a small jet plane

Marsh, was significantly bigger. It had a wingspan that ranged from 9 to 20 feet (2.7 to 6 meters), according to a 2000 study in Current Research in Earth Sciences, a peer-reviewed bulletin of the Kansas Geological Survey.

The smallest pterosaur, called Nemicolopterus crypticus, was discovered in the western part of China's Liaoning Province. It had a wingspan of only 10 inches (25 centimeters), according to a description of the animal, published in 2008 in the journal Proceedings of the National Academy of Sciences.

One of the largest pterosaurs is believed to be Quetzalcoatlus northropi, whose wingspan reached 36 feet (11 meters), according to the 2010 PLOS ONE article.

Another large pterosaur was Coloborhynchus capito, which had a wingspan of about 23 feet (seven meters). This discovery, described in a 2012 article in the journal Cretaceous Research, followed an examination of a fossil that had been in the Natural History Museum of London since 1884.

PHYSICAL CHARACTERISTICS

Given the large number of types of pterosaurs, the characteristics of the winged reptiles varied widely depending on the genera.

Pterosaurs often had long necks, which sometimes had throat pouches similar to pelicans' for catching fish. Most pterosaur skulls were long and full of needle-like teeth. However, pterosaurs of the taxonomic family Azhdarchidae, which ruled the Late Cretaceous skies and included Quetzalcoatlus northropi, were toothless, according to a 2014 study in the journal ZooKeys. ▶

Images: Alamy, Getty Images (life size model)

Pterosaurs were carnivores, though some may have occasionally eaten fruits

Pterosaur's wing shape allowed it to travel long distances without flapping its wings

A distinguishing feature of pterosaurs was the crest on their heads. Though it was initially thought that pterosaurs had no crests, it is now known that crests were widespread across pterosaur genera and came in a variety of different forms.

For instance, some pterosaurs had big, bony crests, while other crests were fleshy with no underlying bone. Some pterosaurs even appear to have had a sail-like crest made up of a membrane sheet connecting two

large bones on the head. "We now know that pterosaur crests had all kinds of [bone and flesh] combinations," Hone told us.

Over the years, scientists have proposed many possible purposes for these crests, including that they were used for heat regulation or to serve as rudders during flight. "But almost all of the hypotheses have failed the most basic tests," Hone said, adding that models show the crests aren't effective rudders and many small pterosaurs have crests even though they wouldn't have needed them to dissipate heat.

What seems most likely is that the crests were used for sexual selection, Hone and his colleagues argued in a 2011 study in the journal Lethaia.

There are several lines of evidence that support this function of the crests, Hone explained. Perhaps the most notable is that juveniles — which look like miniature versions of adult pterosaurs — don't have crests, suggesting the structures are used for something that is only relevant to adults, such as mating.

WHAT DID PTEROSAURS EAT?

Pterosaurs were carnivores, though some may have occasionally eaten fruits. What the reptiles ate

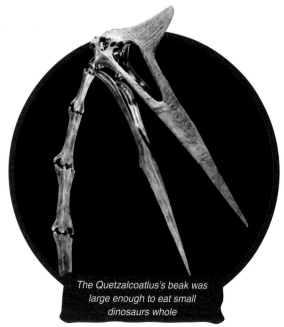

The Quetzalcoatlus's beak was large enough to eat small dinosaurs whole

Some pterosaurs like Geosternbergia hunted small fish, squid and other sea life

depended on where they lived — some species spent their lives around water, while others were more terrestrial.

Terrestrial pterosaurs ate carcasses, baby dinosaurs, lizards, eggs, insects and various other animals. "They were probably fairly active hunters of small prey," Hone said. Water-loving pterosaurs ate a variety of marine life, including fish, squid, crab and other shellfish.

In 2014, Hone sought to learn more about the lives of marine pterosaurs. With these animals, juveniles dominate the fossil record, Hone said. This is odd because young animals are generally those that are targeted by predators, which prevents them from becoming a significant part of the fossil record.

One hypothesis to explain this strange occurrence is that the juvenile pterosaurs often died by drowning instead of being eaten. To test this, Hone and his colleague Donald Henderson modeled how well pterosaurs could float on water (like ducks). They found that pterosaurs floated well, but had poor floating postures, in which their heads rested very close to, if not on, the water.

This suggests that aquatic pterosaurs wouldn't spend much time on the water's surface and would launch into the air shortly after diving for food to avoid drowning. However, young pterosaurs that didn't yet have strong muscles or were still learning to fly would have had more difficulties launching back into the air from a dive, possibly resulting in drowning, Hone said.

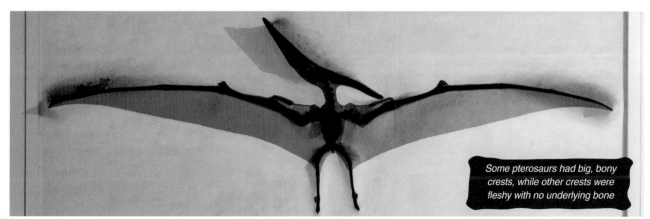

Some pterosaurs had big, bony crests, while other crests were fleshy with no underlying bone

DIPLODOCUS

Diplodocus lived during the late Jurassic Period, about 155 million to 145 million years ago

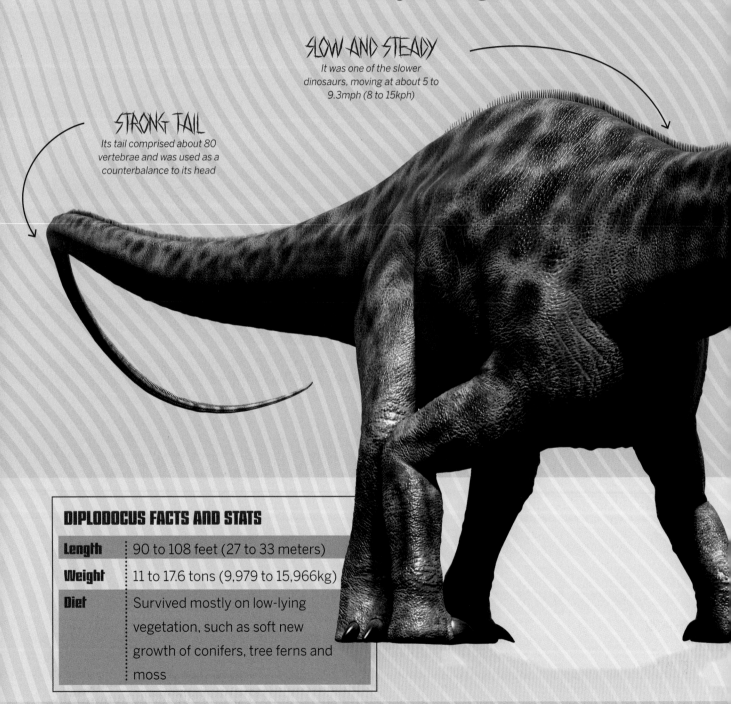

SLOW AND STEADY

It was one of the slower dinosaurs, moving at about 5 to 9.3mph (8 to 15kph)

STRONG TAIL

Its tail comprised about 80 vertebrae and was used as a counterbalance to its head

DIPLODOCUS FACTS AND STATS

Length	90 to 108 feet (27 to 33 meters)
Weight	11 to 17.6 tons (9,979 to 15,966kg)
Diet	Survived mostly on low-lying vegetation, such as soft new growth of conifers, tree ferns and moss

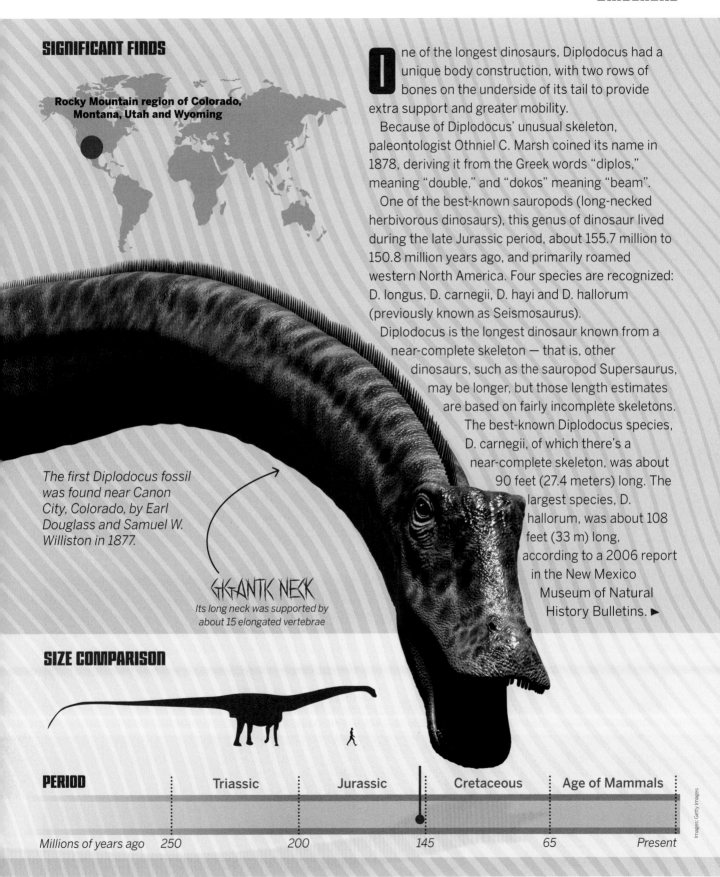

SIGNIFICANT FINDS

Rocky Mountain region of Colorado, Montana, Utah and Wyoming

One of the longest dinosaurs, Diplodocus had a unique body construction, with two rows of bones on the underside of its tail to provide extra support and greater mobility.

Because of Diplodocus' unusual skeleton, paleontologist Othniel C. Marsh coined its name in 1878, deriving it from the Greek words "diplos," meaning "double," and "dokos" meaning "beam".

One of the best-known sauropods (long-necked herbivorous dinosaurs), this genus of dinosaur lived during the late Jurassic period, about 155.7 million to 150.8 million years ago, and primarily roamed western North America. Four species are recognized: D. longus, D. carnegii, D. hayi and D. hallorum (previously known as Seismosaurus).

Diplodocus is the longest dinosaur known from a near-complete skeleton — that is, other dinosaurs, such as the sauropod Supersaurus, may be longer, but those length estimates are based on fairly incomplete skeletons. The best-known Diplodocus species, D. carnegii, of which there's a near-complete skeleton, was about 90 feet (27.4 meters) long. The largest species, D. hallorum, was about 108 feet (33 m) long, according to a 2006 report in the New Mexico Museum of Natural History Bulletins. ▶

The first Diplodocus fossil was found near Canon City, Colorado, by Earl Douglass and Samuel W. Williston in 1877.

GIGANTIC NECK
Its long neck was supported by about 15 elongated vertebrae

SIZE COMPARISON

PERIOD	Triassic	Jurassic	Cretaceous	Age of Mammals

Millions of years ago   250   200   145   65   *Present*

The majority of Diplodocus' length was taken up by its neck and tail. For instance, the neck alone of D. carnegii was at least 21 feet (6.4 meters) long, according to a 2011 study in the Journal of Zoology, and its tail was even longer.

Diplodocus' long tail possibly served as a counterbalance to its neck. A 1997 study in the journal Paleontology also found that diplodocids — dinosaurs in the Diplodocidae taxonomy family, which includes Diplodocus and Apatosaurus — could whip the tips of their tails at supersonic speeds, producing a cannon-like boom, possibly to intimidate would-be attackers or rivals, or for the purposes of communication and courtship.

MASSIVE DINOSAUR

Estimating the mass of dinosaurs is often difficult, and modern estimates of Diplodocus' mass

(excluding D. hallorum) have ranged between 11 and 17.6 tons (10 to 16 metric tonnes). The dinosaur's large tail placed its center of mass pretty far back on its body, said David Button, a paleontologist at the University of Bristol in the United Kingdom.

"It seems that its center of mass is so far back that it wouldn't have been able to walk very quickly," Button told us, adding that this center-of-mass position would have also made rearing up on its hind legs rather easy for Diplodocus.

Based on a 1910 reconstruction of Diplodocus by paleontologist Oliver P. Hay, scientists initially thought that Diplodocus' posture was more lizard-like, with splayed limbs. However, paleontologist William J. Holland argued that such a posture would have required a large ditch to accommodate the dinosaur's stomach. In the 1930s, fossil footprints,

A Diplodocus' neck alone could be up to 21 feet long. It needed a long tail to stay stable

Dippy the Diplodocus on display at the National History Museum in London, UK

Diplodocus is the most displayed dinosaur, with skeletons shown in museums around the world

> ❝ *Diplodocus walked with its broad legs straight down, like an elephant* ❞

or "trackways," suggested Diplodocus walked with its broad legs straight down, like an elephant.

Like some other sauropods, Diplodocus' nasal openings sat high up on its forehead instead of at the end of its snout. At one point, scientists thought that Diplodocus might have had a trunk. However, a 2006 study in the journal Geobios concluded that Diplodocus didn't have neuroanatomy that would be able to support a trunk, after comparing skulls of the dinosaur with those of elephants.

Another theory explaining Diplodocus' high nasal openings proposed the dinosaur needed this adaptation to live in water. But sauropods likely weren't suited for aquatic life, because they had pockets of air inside their bodies that would have made them too buoyant — and unstable — in deep water, according to a 2004 study in the journal Biology Letters.

Diplodocus had five-toed broad feet, with the "thumb" toes sporting a claw that was unusually large compared with other sauropods. It's not known what

purpose this claw served for Diplodocus or other sauropods.

Fossilized skin impressions described in a 1992 Geology paper suggest that diplodocids may have had small, keratinous spines along their tails, bodies and necks.

Like other sauropods, Diplodocus probably grew very quickly, reached sexual maturity at about ten years of age, and continued to grow throughout life, according to a 2004 study in the journal Organisms Diversity & Evolution.

No direct evidence of Diplodocus' nesting habits exists, but it's possible the dinosaur, similar to other sauropods, laid its eggs in a communal area containing vegetation-covered shallow pits.

WHAT DID DIPLODOCUS EAT?

According to a 2009 article in the journal Acta Palaeontologica Polonica, Diplodocus probably held its neck at a 45-degree angle most of the time. ▶

Images: Getty Images, Alamy (fossils)

The dinosaur's teeth were bunched at the front and would strip foliage from branches

Its tail could make a booming sound when whipped to intimidate potential predators

> " *Diplodocus had a very flexible neck, contrary to some previous research* "

However, it's unclear if the animal had the neck flexibility that would allow it to reach both plants on the ground and leaves at the top of trees without moving its body.

"Neck flexibility is a controversial topic in sauropods," Button said. Most recently, a 2014 study in the journal PeerJ suggests Diplodocus had a very flexible neck, contrary to some previous research.

But even if the dinosaur couldn't lift its head up very high, it could still rear up on its hind legs to reach the top of tall trees, Button notes. "It wouldn't have had much trouble low browsing and high browsing," he said.

Diplodocus had a number of small, forward-pointing, peg-like teeth that were bunched in the front of its mouth. The teeth were slender and delicate, and replaced very quickly, Button said.

According to a 2013 study in the journal PLOS ONE, Diplodocus had a tooth-replacement rate of one tooth every 35 days, while the sauropod Camarasaurus, which lived in the same areas at the same time as Diplodocus, replaced one tooth every 62 days. The high tooth-replacement rate of Diplodocus suggests the animal was eating abrasive food, such as soft plants that contained silica or grit-covered plants on the ground, Button said.

In 2012, Button and his colleagues sought to learn what, exactly, Diplodocus could eat. They modeled the mechanical stresses the animal would have experienced under normal biting, branch-stripping and bark-stripping behaviors, and found the animal wouldn't have been able to handle stripping bark from a tree, according to their study published in the journal Naturwissenschaften.

In a follow-up study, the researchers used cranial biomechanical models to further investigate the dinosaurs' feeding habits, and determine how it could coexist with Camarasaurus when both animals required a lot of sustenance and lived in a relatively resource-poor environment.

"Our main finding was that compared to Camarasaurus, Diplodocus had a weaker overall bite force," Button said. "It used very different jaw muscles that emphasized a horizontal rather than vertical movement, or sliding instead of hard biting."▶

Diplodocus' jaws were suited for grabbing and pulling soft vegetation

This means that the two animals engaged in niche partitioning — they ate two completely different foods. Camarasaurus' skull and jaw were adapted to accommodating high stresses, allowing it to eat tough leaves and branches. Diplodocus, on the other hand, was more suited to eating ferns and stripping soft leaves off trees, Button said.

And rather than chewing, Diplodocus spent a considerable time fermenting its food in its expanded gut, and probably didn't use stones to help with digestion, Button said.

FOSSIL FINDS

The first Diplodocus fossil was found near Cañon City, Colorado, by Benjamin Mudge and Samuel W. Williston in 1877, and was named by Marsh in 1878.

A number of Diplodocus fossils have been found in the Rocky Mountain region of Colorado, Montana, Utah and Wyoming, areas that are all a part of the fossil-rich Morrison Formation.

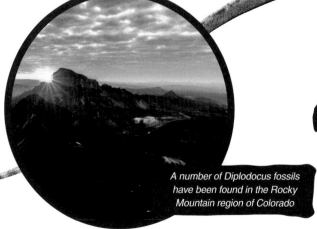

A number of Diplodocus fossils have been found in the Rocky Mountain region of Colorado

66 *Diplodocus had a weaker bite force than Camarasaurus* 99

Diplodocus may have had keratinous spines running from its neck and along its tail

Thanks to steel magnate Andrew Carnegie, who donated casts of complete skeletons to various European monarchs, Diplodocus is among the most displayed dinosaurs in the world. Diplodocus can be viewed at a number of museums globally, including the Carnegie Museum of Natural History in Pittsburgh and the Paris Museum of Natural History.

In 2015, London's Natural History Museum announced that its iconic Dippy — a replica of the near-complete D. carnegii fossil discovered in 1898 — would be going on a national tour, and that it would be replaced with a model of a blue whale during its travels.

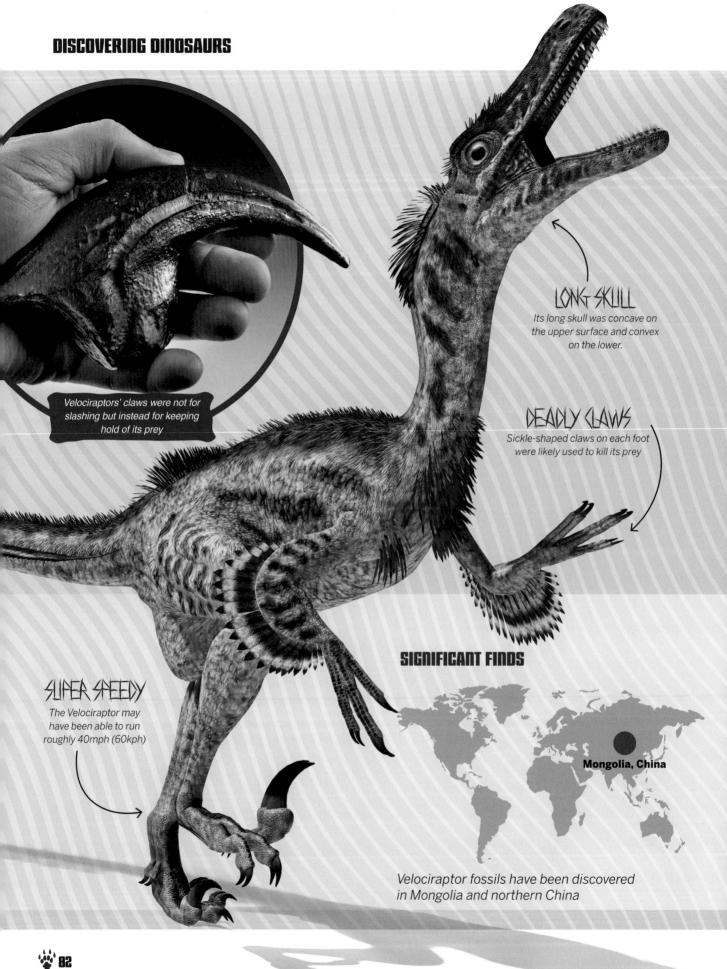

Velociraptors' claws were not for slashing but instead for keeping hold of its prey

LONG SKULL

Its long skull was concave on the upper surface and convex on the lower.

DEADLY CLAWS

Sickle-shaped claws on each foot were likely used to kill its prey

SUPER SPEEDY

The Velociraptor may have been able to run roughly 40mph (60kph)

SIGNIFICANT FINDS

Mongolia, China

Velociraptor fossils have been discovered in Mongolia and northern China

VELOCIRAPTOR

These small, bird-like dinosaurs roamed the Earth 75 million to 71 million years ago, toward the end of the Cretaceous period

In 1924, Henry Fairfield Osborn, then-president of the American Museum of Natural History, named the Velociraptor. He derived its name from the Latin words "velox" (swift) and "raptor" (robber or plunderer), as an apt description of its agility and carnivorous diet.

Earlier that year, Osborn had called the dinosaur Ovoraptor djadochtari in an article in the popular press, but the creature wasn't formally described in the article and the name "Ovoraptor" wasn't mentioned in a scientific journal, making Velociraptor the accepted name.

There are two Velociraptor species, V. mongoliensis and V. osmolskae, the second of which was only identified relatively recently, in 2008.

A member of the Dromaeosauridae family of small- to medium-sized bird-like dinosaurs, Velociraptor was roughly the size of a small turkey and smaller than others in this family of dinosaurs, which included Deinonychus and Achillobator. Adult Velociraptors grew up to 6.8 feet (two meters) long, 1.6 feet (0.5 meters) tall at the hip and weighed up to 33 pounds (15 kilograms).

Like Tyrannosaurus rex, Velociraptor played a prominent role in the *Jurassic Park* movies, but scientists do not believe it resembled anything close to its Hollywood depiction in terms of size or appearance. In fact, the movies' Velociraptor was actually modeled after Deinonychus, and sported a similar size and snout. While the Velociraptor was featherless in the ▶

VELOCIRAPTOR FACTS AND STATS

Length	Up to 6.8 feet (two meters)
Height	1.6 feet (0.5 meters) at the hips
Weight	Up to 33 pounds (15kg)
Diet	Meat; survived on mostly small animals, such as reptiles, amphibians and smaller, slower dinosaurs

SIZE COMPARISON

PERIOD		Triassic	Jurassic	Cretaceous	Age of Mammals
Millions of years ago	250	200	145	65	Present

Velociraptors had feathers on their arms but they would not have been used for flying

Widely separated and serrated teeth allowed the dinosaur to bite and hold onto prey

> " *Velociraptor's tail of hard, fused bone was inflexible, but likely kept it balanced as it ran, hunted and jumped* "

movies, paleontologists discovered quill knobs (places where the flight-related feathers of birds are anchored to the bone) on a well-preserved Velociraptor forearm from Mongolia in 2007, indicating the dinosaur was likely to have had feathers.

Despite having feathers, however, the arms of Velociraptors were too short to allow them to fly or even glide. The find suggests that the dinosaurs' dromaeosaurid ancestors could fly at one point, but lost that ability, according to the study published in the journal Science.

Velociraptor retained its feathers, and possibly used them to attract mates, regulate body temperature, protect eggs from the environment or generate thrust and speed while running up inclines.

Velociraptor had a relatively large skull, which was about 9.1 inches (23 centimeters) long, concave on the upper surface and convex on the lower surface, according to a 1999 description of a Velociraptor skull, published in the journal Acta Palaeontologica Polonica. Additionally, its snout was long, narrow and shallow, and made up about 60 percent of the dinosaur's entire skull length.

Velociraptor had 13 to 15 teeth in its upper jaw

They had to be fast to chase and hunt small mammals and other dinosaurs

and 14 to 15 teeth in its lower jaw. These teeth were widely spaced and serrated, though more strongly on the back edge than on the front.

Velociraptor's tail of hard, fused bones was inflexible, but likely kept the dinosaur balanced as it ran, hunted and jumped.

Velociraptor, like other dromaeosaurids, had two large hand-like appendages with three curved claws. They also had a sickle-shaped talon on the second toe of each foot. They normally kept these talons off the ground like folded switchblades, and used them as hooks to keep their prey from escaping (similar to modern birds of prey), according to a study published in 2011 in the journal PLOS ONE.

WHAT DID VELOCIRAPTOR EAT?

Velociraptor was a carnivore that hunted and scavenged for food. "It spent the vast majority of the time eating small things," which likely included reptiles, amphibians, insects, small dinosaurs and mammals, said David Hone, a paleontologist at Queen Mary University of London.

The fast predator also appears to have had a complicated relationship with Protoceratops, a sheep-sized herbivore and ancestor to Triceratops. In 1971, a Polish-Mongolian team discovered the famous "Fighting Dinosaurs" specimen — fossils of a Velociraptor ▶

Images: Getty Images, Alamy (above)

The dinosaurs could run at speeds of up to 40mph, using their feathers to generate thrust

> *An attack on such a large animal [as Protoceratops] probably wasn't common*

Evidence shows Velociraptors hunted Protoceratops, but it may not have been common

Череп протоцератопса
Protoceratops andrewsi
Granger et Gregory
Поздний мел, Монголия

An artist's rendition of a Velociraptor and Protoceratops engaged in combat

Despite often being depicted in media as pack hunters, Velociraptors hunted alone

and Protoceratops locked in a death grip, in which the Velociraptor had embedded one of its foot claws into the neck of the Protoceratops while the Protoceratops had bitten down on (and probably broken) one of the Velociraptor's arms.

Preserved in sand deposits after being buried by a collapsing sand dune or sudden sandstorm, the pair proved that Velociraptors hunted for food, but an attack on such a large animal probably wasn't common. "Few predators ever take on prey bigger than 50 percent of their body mass," Hone told us, adding that the Velociraptor could have been starving or simply "young and dumb".

But that's not to say Velociraptor didn't frequently eat Protoceratops carcasses. In 2008, researchers unearthed Protoceratops fossils marred with marks and grooves matching raptor teeth, as well as two teeth that belonged either to Velociraptor or another dromaeosaurid.

After analyzing the remains, Hone and his colleagues determined that the raptor didn't kill the plant-eater. Instead, it fed on the Protoceratops, which likely had little meat left on it (hence the bite marks on the herbivore's jaws and raptor's knocked-out teeth), according to the study, which was published in 2010 in the journal Palaeogeography, Palaeoclimatology, Palaeoecology.

In 2012, Hone and his colleagues also discovered that Velociraptors sometimes ate pterosaurs, when the team found a large pterosaur bone in the guts of a Velociraptor. The pterosaur had a wingspan of about 6.5 feet (2 meters) and would have been a formidable foe even if it were sick and injured, suggesting the Velociraptor most likely scavenged the pterosaur bone, Hone said.

FOSSIL DISCOVERIES

The first Velociraptor fossil was discovered by Peter Kaisen on the first American Museum of Natural History expedition to the Outer Mongolian Gobi Desert in August 1923. The fossil consisted of a skull that was crushed but complete and a toe claw.

Further Velociraptor fossils have been found in the Gobi Desert, which covers southern Mongolia and parts of northern China. Velociraptor mongoliensis has only been discovered in the Djadochta (Djadokhta) Formation, which is in the Mongolian province of Ömnögovi.

Velociraptor osmolskae was discovered at the Bayan Mandahu Formation in Inner Mongolia, China. The species was described based on a partial adult skull.

Like the "Fighting Dinosaurs," other Velociraptor fossils were found in arid, sand dune environments.

TRIVIA

Everything you need to know about
the amazing world of dinosaurs

Images: Getty Images, Alamy (grit & Giganotosaurus), Shutterstock (T Rex modern world)

> *Had the asteroid hit just about anywhere else on the planet, dinosaurs might still roam the Earth*

90

112

HOW DID DINOSAURS GET SO BIG?

Dinosaurs dwarfed some of the largest animals alive today

The secret to mega-dinosaurs' impressive sizes may be that the reptiles used more of their energy for growing and less for keeping their bodies warm compared with some creatures. A new model could help explain how some dinosaurs, such as long-necked sauropods, could have achieved masses of around 60 tons — about eight times the mass of an African elephant, the largest land animal alive today.

The two main factors that determine vertebrate size are the amount of available food and how the creature expends its energy, said researcher Brian K. McNab, a paleontologist at the University of Florida. For example, elephants can be quite large because they feed off grasses, a relatively abundant food supply as opposed to, say, the nectar that hummingbirds and bees consume, McNab said.

Energy expenditure depends in part on how an organism controls its body temperature. Mammals and birds, which are warm-blooded, must expend energy to keep their internal body temperatures constant, and so they have a high metabolic rate. But cold-blooded creatures such as reptiles rely on their environment for body heat, and their internal temperature fluctuates depending on the surrounding conditions.

Warm-blooded animals must therefore eat a lot more than cold-blooded animals to produce their own body heat.

Whether dinosaurs were warm-blooded or cold-blooded has been a hotly debated issue among paleontologists. McNab attempted to answer this question by looking at what food resources were ▶

These reptiles used more of their energy for growing and less for keeping their bodies warm

SPINOSAURUS

At up to 59 feet long and weighing up to 20 tons, the Spinosaurus is the largest meat-eating dinosaur known to have existed.

The exact size and weight of Giganotosaurus is unknown due to incomplete skeletons

GIGANOTOSAURUS

" It may have been possible for dinosaurs to be warm-blooded "

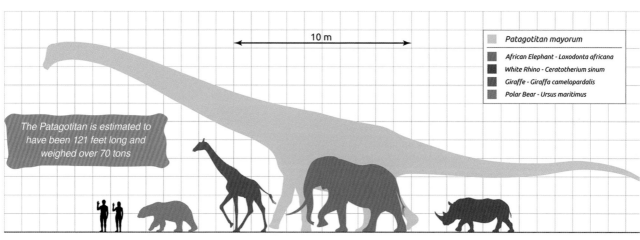

10 m

Patagotitan mayorum
African Elephant - Loxodonta africana
White Rhino - Ceratotherium sinum
Giraffe - Giraffa camelopardalis
Polar Bear - Ursus maritimus

The Patagotitan is estimated to have been 121 feet long and weighed over 70 tons

available to dinosaurs, and included this factor in his model that describes how vertebrate size, energy expenditure and food resources tie together.

If resources were much more abundant in the Mesozoic era — the time period when the dinosaurs lived — than today, it may have been possible for dinosaurs to be warm-blooded, as there would have been sufficient food to maintain their body temperature. Indeed, blue whales, the largest creatures thought to have ever lived on Earth, are warm-blooded. They fuel their 160-ton bodies by feeding off of the plentiful resources in marine environments.

However, McNab concluded this was not the case for dinosaurs. "I think it was impossible for [dinosaurs] to have really high metabolic rates like mammals and birds, simply because the resources weren't there," he told us.

For example, there were no grasses in the Mesozoic, which are a major food source for most herbivores, McNab said.

"How is it that dinosaurs got larger than mammals if the resources were either equal to or poorer than today? My argument is it's because they took most of the energy they consumed and put it into growth rather than into maintenance of a high body temperature," he said.

So were dinosaurs cold-blooded? Not exactly, said McNab. He thinks that dinosaurs were "homeothermic," somewhere in between warm and cold-blooded. They did not have a high metabolic rate, but their internal temperature did not fluctuate like that of cold-blooded creatures. Instead, their sheer size kept their body temperature at a constant level.

"When you're that big, you can't cool off rapidly like a small lizard can," explained McNab. "You have a large volume, and you have comparatively small surface area. And so if you're warm, you'll continue to stay warm, unless something unforeseen happens."

> **When you're that big, you can't cool off rapidly like a small lizard can**

DIPLODICUS

If warm-blooded, these giants would need to eat a lot of food to maintain their body temperature

SHRINKING DINOSAURS EVOLVED INTO FLYING BIRDS

EUDIMORPHODON

How did theropods become the animals we know today?

" *Birds are the only dinosaurs that are still alive today* "

Quetzalcoatlus were the giants of the sky during the Cretaceous period. They could grow to the size of a small plane

Today's birds evolved from dinosaurs that shrank continuously for 50 million years, researchers say.

Dinosaurs first emerged about 245 million years ago and they came to dominate the planet. They included the largest animals ever to walk across the surface of Earth. Their reign came to an end with a bang — a cosmic impact from an asteroid or comet maybe six miles (ten kilometers) wide. However, not all of the dinosaurs were killed by this impact — some became the birds now found on every continent on Earth.

"Birds are the only dinosaurs that are still alive today," said Michael Lee, an evolutionary biologist at the South Australian Museum and the University of Adelaide in Australia.

Birds evolved from carnivorous dinosaurs known as theropods, which included giant predators such as Tyrannosaurus rex and talon-footed raptors such as Deinonychus. Recent studies of theropods keep finding more and more bird-like traits, such as feathers, wishbones, hollow skeletons and three-fingered hands, Lee said. ▶

Hummingbirds evolved from massive ground-dwelling theropods

ARCHAEOPTERYX

Images: Getty Images

A group of Pterodactylus fly though the sky. Winged dinosaurs first appeared in the late Triassic period and roamed the skies until the end of the Cretaceous period

To learn more about how small, graceful fliers such as hummingbirds evolved from massive ground-dwelling theropods, scientists developed a detailed family tree of birds and their dinosaur ancestors, mapping out this evolutionary transformation. They analyzed more than 1,500 anatomical traits from 120 species of early birds and all branches of the theropod family tree, and estimated when the two diverged from each other and how quickly each lineage changed.

The researchers discovered that the branch of theropods that gave rise to modern birds consisted of the only theropods that kept shrinking. In the lineage that led directly to birds, body size decreased during the course of at least 12

Birds were able to survive the meteorite impact event that wiped out the dinosaurs

ANHANGUERA

PTERNOSAURUS

different stages over 50 million years or so, from stiff-tailed carnivores known as tetanurans that lived about 198 million years ago and weighed about 359 pounds (163 kilograms) to Archaeopteryx, long considered the first known bird, which lived about 150 million years ago and weighed an average of about 1.7 pounds (0.8 kilograms).

"No other dinosaur group underwent such an extended period of shrinkage," Lee told us.

In addition to this miniaturization, the scientists discovered that bird precursors evolved new skeletal adaptations — such as wings, enlarged brains, smaller teeth and larger eyes — four times faster than other dinosaurs. "Birds arose from the most 'evolvable' dinosaurs," explained Lee.

All in all, "being smaller and lighter in the land of giants, with rapidly evolving anatomical adaptations, provided these bird ancestors with new ecological opportunities, such as the ability to climb trees, glide and fly," Lee said. "This evolutionary flexibility helped birds survive the deadly meteorite impact, which killed off all their dinosaurian cousins."

Ultimately, "birds out-shrank and out-evolved their dinosaurian ancestors, surviving where their larger, less-evolvable relatives could not," Lee continued.

In the future, the researchers would like "to see these methods applied to help understand rates and patterns of evolution in other groups," Lee said. "For instance, how fast did mammals evolve after the dinosaur extinctions? We know they evolved rapidly, but we don't have exact rates yet."

> " This evolutionary flexibility helped birds survive the deadly meteorite impact "

Birds out-shrank and out-evolved their dinosaurian cousins

"These herbivores had by far the longest necks of any known animal"

Titanosaurs were a family of sauropods that included the largest land animals known to have existed, such as the Argentinosaurus

BRONTOMERUS

HOW DINOSAURS GREW THE WORLD'S LONGEST NECKS

How did the largest of all dinosaurs evolve necks longer than any other creature that has ever lived?

The largest creatures to ever walk the Earth were the long-necked, long-tailed dinosaurs known as the sauropods. These herbivores had by far the longest necks of any known animal. The dinosaurs' necks reached up to 50 feet (15 meters) in length, six times longer than that of the current world-record holder, the giraffe, and at least five times longer than those of any other animal that has lived on land.

"They were really stupidly, absurdly oversized," said researcher Michael Taylor, a vertebrate paleontologist at the University of Bristol in England. "In our feeble, modern world, we're used to thinking of elephants as big, but sauropods reached ten times the size elephants do. They were the size of walking whales."

AMAZING NECKS

To find out how sauropod necks got so long, scientists analyzed other long-necked creatures and compared sauropod anatomy with their nearest living relatives: birds and crocodilians. ►

"Extinct animals — and living animals, too, for that matter — are much more amazing than we realize," Taylor told us. "Time and again, people have proposed limits to possible animal sizes, like the 16-foot (five-meter) wingspan that was supposed to be the limit for flying animals. And time and again, they've been blown away. We now know of flying pterosaurs with 33-foot (ten-meter) wingspans. And these extremes are achieved by a startling array of anatomical innovations."

Animals today such as the giraffe and ostrich can't match the neck-length of sauropods

The longest marine reptile's neck belonged to the plesiosaurs at 23 feet

Among living animals, adult bull giraffes have the longest necks, capable of reaching about eight feet (2.4 meters) long. No other living creature exceeds half this length. For instance, ostriches typically have necks only about three feet (one meter) long. When it comes to extinct animals, the largest land-living mammal of all time was the rhino-like creature Paraceratherium, which had a neck maybe 8.2 feet (2.5 meters) long. The flying reptiles known as pterosaurs could also have surprisingly long necks, such as Arambourgiania, whose neck may have exceeded ten feet (three meters).

The necks of the Loch Ness Monster-like marine reptiles known as plesiosaurs could reach an impressive 23 feet (seven meters), probably because the aquatic environment they lived in could better support their weight. But these necks were still less than half the lengths of the longest-necked sauropods.

SAUROPOD SECRETS

In their study, Taylor and his colleagues found that the neck bones of sauropods possessed a number of traits that supported such long necks. For instance, air often made up 60 percent of these animals' neck bones, with some as light as birds' bones, making it easier to support long chains of them. The muscles, tendons and ligaments were also positioned around these vertebrae in a way that

MASSOSPONDYLUS

> "Sauropods possessed a number of traits that supported such long necks"

Dinosaurs such as Omeisaurus needed long tails to counterbalance their necks

helped maximize leverage, making neck movements more efficient.

In addition, the dinosaurs' giant torsos and four-legged stances helped provide a stable platform for their necks. In contrast, giraffes have relatively small torsos, while ostriches have two-legged stances.

Sauropods also had plenty of neck vertebrae, up to 19. In contrast, nearly all mammals have no more than seven, from mice to whales to giraffes, limiting how long their necks can get. (The only exceptions among mammals are sloths ▶ and aquatic mammals known as sirenians, such as manatees.)

Moreover, while pterosaur Arambourgiania had a relatively giant head with long, spear-like jaws that it likely used to help capture prey, sauropods had small, light heads that were easy to support. These dinosaurs did not chew their meals, lacking even cheeks to store food in their mouths; they merely swallowed it, letting their guts break it down.

"Sauropod heads are essentially all mouth. The

PARACERATHERIUM

jaw joint is at the very back of the skull, and they didn't have cheeks, so they came pretty close to having Pac Man/Cookie Monster flip-top heads," researcher Mathew Wedel at the Western University of Health Sciences in Pomona, California, told us.

"It's natural to wonder if the lack of chewing didn't, well, come back to bite them, in terms of digestive efficiency. But some recent work on digestion in large animals has shown that after about three days, animals have gotten all the nutrition they can from their food, regardless of particle size.

"And sauropods were so big that the food would have spent that long going through them anyway," explained Wedel. "They could stop chewing entirely, with no loss of digestive efficiency."

WHAT'S A LONG NECK GOOD FOR?

Furthermore, sauropods and other dinosaurs probably could breathe like birds, drawing fresh air through their lungs continuously, instead of having to breathe out before breathing in to fill their lungs with fresh air like mammals do. This may have helped sauropods get vital oxygen down their long necks to their lungs.

"The problem of breathing through a long tube is something that's very hard for mammals to do. Just try it with a length of garden hose," Taylor said.

As to why sauropods evolved such long necks, there are currently three theories. Some of the dinosaurs may have used their long necks to feed on high leaves, like giraffes do. Others may have used their necks to graze on large swaths of vegetation by sweeping the ground from side to side like geese do. This helped them make the most out of every step, each of which would be a big deal for such heavy creatures.

Scientists have also suggested that long necks may have been sexually attractive, therefore driving the evolution of ever-longer necks; however, Taylor and his team have found no evidence this was the case.

In the future, the researchers plan to delve even deeper into the mysteries of sauropod necks. For instance, Apatosaurus had

Long necks would help with reaching food, but they could have had other uses

APATOSAURUS

Images: Getty Images, Alamy (skeleton)

"really sensationally strange neck vertebrae," Taylor said. The scientists suspect the necks of Apatosaurus were used for "combat between males — fighting over females, of course."

> *Sauropods had small, light heads that were easy to support*

Sauropods had more than double the neck vertebrae of most mammals

COULD EVOLUTION EVER BRING BACK THE DINOSAURS?

Could the evolutionary process make Jurassic Park a reality?

Did you watch the 1993 movie blockbuster *Jurassic Park* and wonder, "Could this happen for real? Could the dinosaurs ever come back?" The idea that these mighty creatures could wander our Earth again some day is for most humans both fascinating and terrifying in equal measure.

Even real-life scientists are intrigued as to whether the evolutionary process could bring us back to the time of the Tyrannosaurs. But Susie Maidment, a vertebrate paleontologist at London's Natural History Museum, quickly dismissed the notion that a DNA-filled mosquito preserved in amber for millions of years — as in *Jurassic Park* — could ever help recreate an extinct dinosaur.

"We do have mosquitos and biting flies from the

time of the dinosaurs, and they do preserve in amber," Maidment said. "But when amber preserves things, it tends to preserve the husk, not the soft tissues. So, you don't get blood preserved inside mosquitos in amber."

Researchers have found blood vessels and collagen in dinosaur fossils, but these components don't have actual dinosaur DNA in them. Unlike collagen or other robust proteins, DNA is very fragile, and sensitive to the effects of sunlight and water. The oldest DNA in the fossil record is around 1 million years old, and the dinosaurs died out about 66 million years ago.

Maidment added: "Although we have what appears to be blood from mosquitos up to 50 million years old, we haven't found DNA, and in order to reconstruct ▶

Jurassic Park *and its sequels led to people imagining the possibility of dinosaurs living in our world*

Image: Shutterstock

Ichthyosaurus were marine reptiles that lived during the late Triassic/early Jurassic period

> *They say that dinosaurs went extinct, but only the non-avian dinosaurs went extinct*

might disrupt our genomes, our physiology and behavior beyond our control," he told us. This, in turn, could create the right conditions for evolution to take a path toward reinventing the ancient reptiles.

However, while evolution might not be directional in any particular sense, something we do know is that we don't see the same animal evolving again, Maidment countered. "We can see an animal that is closely related occupying a similar ecological niche — for example, ichthyosaurs were marine reptiles with long pointy snouts and dolphin-like body shapes and tails," she explained. "Today we see the dolphin, and they probably occupy a similar ecological niche. But we wouldn't describe a dolphin as an ichthyosaur because they don't possess the anatomical characteristics that allow them to be ichthyosaurs."

Besides, dinosaurs never quite died out in the first place, Maidment said. Birds evolved from meat-eating dinosaurs, and thus in strict biological

something, we need DNA."

Jamal Nasir, a geneticist at the University of Northampton in the United Kingdom, said he wouldn't rule out the idea of dinosaurs evolving back from the dead. In his opinion, evolution isn't fixed or planned. In other words, anything could happen. "Evolution is largely stochastic [randomly determined], and evolution doesn't necessarily have to go in a forward direction; it could have multiple directions. I would argue that going back to dinosaurs is more likely to happen in reverse, because the building blocks are already there."

Of course, Nasir pointed out, the right conditions would have to exist for dinosaurs to reappear. "Clearly, one could imagine viral pandemics that

UTAHRAPTOR

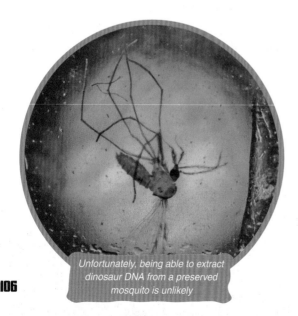

Unfortunately, being able to extract dinosaur DNA from a preserved mosquito is unlikely

Despite some visual similarities, ichthyosaur and dolphins don't share anatomical features

definition, everything that evolved from this common ancestor is a dinosaur, sharing the same anatomical characteristics, she said.

"Dinosaurs are still with us," Maidment said. "They say dinosaurs went extinct, but only the non-avian dinosaurs went extinct. Birds are dinosaurs, and birds are still evolving, so we will certainly see new species of birds evolving — and those will be new species of dinosaur."

Some scientists are even dabbling with the evolution process by trying to reverse engineer a chicken into a dinosaur, dubbed the "chickenosaurus." However, this beast, if it ever comes to fruition, would not be a replica of a dinosaur, but rather a modified chicken, Jack Horner, a research associate at the Burke Museum at the University of Washington, told us.

Things have changed drastically over 66 million years, and if one day a dinosaur evolved back onto Earth, it would be to a very different world.

"An animal that died out naturally, perhaps 150 million years ago, is not going to recognize anything in this world if you bring it back," Maidment noted "What is it going to eat when grass hadn't [yet] evolved back then? What is its function, where do we put it, does anyone own it?"

That said, it may be best to let sleeping dinosaurs lie, she said.

Our planet is completely different to how it was when dinosaurs roamed it

COULD DINOSAURS HAVE SURVIVED THE ASTEROID?

What were the odds of avoiding a mass extinction?

The age of dinosaurs met an unlikely end — because had the cosmic impact that doomed it hit just about anywhere else on the planet, the "terrible lizards" might still roam the Earth.

The impact of an asteroid about six miles (ten kilometers) wide about 66 million years ago created a crater more than 110 miles (180 kilometers) across near what is now the town of Chicxulub in Mexico's Yucatán Peninsula. The meteor strike would have released as much energy as 100 trillion tons of TNT, over a billion times more than the atom bombs that destroyed Hiroshima and Nagasaki combined. The blast is ►

The mass extinction event occurred around 66 million years ago, at the end of the Cretaceous period

Image: Alamy

More than three-quarters of all life was wiped out by the extinction event

Volcanic eruptions contributed to climate change and the mass extinction

thought to have ended the age of dinosaurs, killing off more than 75 percent of all land and sea animals.

Prior work suggested the Chicxulub impact would have lofted huge amounts of ash, soot and dust into the atmosphere, choking off the amount of sunlight reaching Earth's surface by as much as 80 percent. This would have caused Earth's surface to rapidly cool, leading to a so-called "impact winter" that would have killed off plants, causing a global collapse of terrestrial and marine food webs.

> **The probability of the mass extinction occurring was only 13 percent**

To explain why the Chicxulub impact winter proved so catastrophic, Japanese scientists previously suggested the super-hot debris from the meteor strike not only caused wildfires across the planet, but also ignited rocks loaded with hydrocarbon molecules like oil. They calculated that such oily rocks would have generated vast amounts of soot.

The amount of hydrocarbons in rocks varies widely depending on location. In a recent study, the Japanese researchers analyzed the places on Earth where an asteroid impact could have happened to cause the level of devastation seen with the Chicxulub event.

The scientists now find the

Had the asteroid landed elsewhere, life as we know it would be completely different

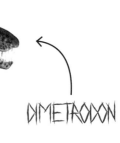

DIMETRODON

Hot debris caused fires and rocks loaded with hydrocarbon to catch fire

asteroid that wiped out the dinosaurs happened to hit an unlucky spot — had it landed in about 87 percent of anywhere else on Earth, the mass extinction might not have occurred.

"The probability of the mass extinction occurring was only 13 percent," said study lead author Kunio Kaiho, a geochemist at Tohoku University in Sendai, Japan.

The scientists ran computer models simulating the amount of soot that asteroid impacts would have generated depending on the amount of hydrocarbons in the ground. They next estimated the climate effects caused by these different impact scenarios.

The researchers calculated the level of climate change needed to cause a mass extinction was a 14.4 to 18 degrees Fahrenheit (8 to 10 degrees Celsius) drop in global average surface air temperatures. This would involve an asteroid impact sending 385 million tons (350 million metric tonnes) of soot into the stratosphere.

The scientists found that a mass extinction would have occurred from the impact only if it had hit 13 percent of the surface of the Earth, including both land and oceans. "If the asteroid had hit a low- to medium-level hydrocarbon area on Earth, occupying approximately 87 percent of the Earth's surface, mass extinction could not have occurred," Kaiho told us.

The scientists also analyzed the level of climate change "caused by large volcanic eruptions that may have contributed to other mass extinctions," Kaiho said, to further understand the processes behind those mass extinctions.

Kaiho and his colleague Naga Oshima at the Meteorological Research Institute in Tsukuba, Japan, detailed their findings online on 9 November 2017 in the journal Scientific Reports.

Those not killed by the crash or ensuing tidal waves would have suffered under acid rain

Images: Getty Images

Dinosaurs died out around 66 million years ago but scientists estimate that their DNA could only have survived one million years

IS IT POSSIBLE TO CLONE A DINOSAUR?

Molecular paleontologist Mary Schweitzer explains whether there is a chance of cloning the prehistoric creatures

Apologies to people keen to revive extinct dinosaurs, but researchers have never recovered the DNA needed for cloning. But, intriguingly, they have found fragments of mystery DNA in dinosaur bone, experts have told us.

It's unknown whether this DNA is dinosaurian, or whether it belongs to other life-forms, such as microbes; non-dinosaurian animals, such as earthworms; or even paleontologists who have worked with these fossils.

"I've found DNA in dinosaur bone," said Mary Schweitzer, a molecular paleontologist at North Carolina State University. "But we did not sequence it — we couldn't recover it, [and] we couldn't characterize it. Whoever it belongs to is a mystery."

It's no surprise that dinosaur remains contain DNA, she said. Bone is partly made up of a mineral called hydroxyapatite, which has a strong affinity for certain biomolecules, including DNA. In fact, researchers often use hydroxyapatite to purify and concentrate DNA in the lab, Schweitzer said.

"That's one of the reasons that I don't work with DNA myself," Schweitzer explained. "It is too prone to contamination and really difficult to interpret."

Instead, Schweitzer analyzes dinosaur fossils for soft tissue, such as the blood vessels that she and her colleagues found in an 80-million-year-old duck-billed dinosaur. But she has still pondered the steps required to clone an extinct dinosaur. Here is the science that it would take to create an actual *Jurassic Park*-style dinosaur, according to molecular experts.

DNA has been discovered in dinosaur bone, but identifying it seems unlikely

> ❝ *We couldn't recover it, [and] we couldn't characterize it. Whoever it belongs to is a mystery* ❞

HOW LONG CAN DNA SURVIVE?

Scientists need DNA to clone dinosaurs, but an organism's DNA starts decaying the moment after that organism dies.

That's because enzymes (from soil microbes, body cells and gut cells) degrade DNA. So does UV radiation. What's more, oxygen and water can chemically alter DNA, causing the strands to break, said Beth Shapiro, an associate professor in the Department of Ecology and Evolutionary Biology at the University of California, Santa Cruz.

"All of these things will break down the DNA into smaller and more degraded pieces, until eventually, there is nothing left," Shapiro explained to us.

The oldest recovered and authenticated DNA from bone belongs to a 700,000-year-old horse from the frozen Klondike gold fields in Yukon, Canada, said Shapiro, who co-wrote a 2013 study on it in the journal Nature.

Still, it's unclear just how long DNA can survive. Scientists have proposed that DNA can survive as long as a million years, but definitely not more than five million or six million years, Schweitzer said. That's woefully short of 65 million years ago, when the asteroid slammed into Earth and killed the non-avian dinosaurs.

However, more experiments are needed to determine how long, and in what conditions, DNA can survive, Schweitzer said.

Moreover, don't expect a *Jurassic Park* twist to work. In the 1993 blockbuster, scientists find dinosaur DNA in an ancient mosquito caught in amber. But amber, it turns out, does not preserve DNA well. Researchers tried to extract DNA from two stingless bees preserved in copal, a precursor of amber, in a 2013 study published in the journal PLOS ONE.

The researchers couldn't find any "convincing evidence for the preservation of ancient DNA" in

Over 66 million years, DNA breaks down due to enzymes and other elements

either of the two copal samples they studied, and they concluded in the study that "DNA is not preserved in this type of material".

They added: "Our results raise further doubts about claims of DNA extraction from fossil insects in amber, many millions of years older than copal."

DINOSAUR DNA?

If researchers choose to study the DNA lurking in dinosaur bone, it will be difficult to say whether it was dinosaurian in nature, the experts said.

"The DNA fragments that were recovered from that horse bone were short (on average 40-ish letters long) and showed characteristic signs of post-mortem damage," Shapiro explained. "But they could be mapped to the genome of a modern horse, and so we know that they were of horse origin."

In contrast, the dinosaurs' living relatives are birds. But birds evolved out of the theropod line — a group of bipedal, largely carnivorous dinosaurs such as Tyrannosaurus rex and Velociraptor. All of the other dinosaur groups — including the hadrosaurs (the duck-billed dinosaurs), the ceratopsians (such as Triceratops), the stegosaurs and the ankylosaurs — do not have any living relatives to compare DNA with.

In addition, any surviving dinosaur DNA will be

Despite what films tell you, dinosaur DNA cannot be preserved in amber

We are unsure of just how long DNA can survive, as more experiments are needed

highly fragmented and badly damaged.

"Here is a key problem with dinosaur DNA," Shapiro said. "I would then have to ask, 'Is this dinosaur DNA, or microbial DNA that got into the dinosaur bone while it was buried?'"

CLONING ADVENTURES

For the sake of argument, let's say that researchers found fully sequenced dinosaur DNA. This means that researchers would have an entire genome, ▶ including the so-called junk DNA and the viral DNA that's incorporated itself into the dinosaur's genetic code. This viral DNA could be a problem, especially if it could infect modern plants and animals, Schweitzer said.

Next, they'd have to find a host organism to help clone the beast. That would likely be a bird. But a mother bird is a far cry from a mother dinosaur, Schweitzer said.

"There's more to developing a vertebrate organism

CORYTHOSAURUS

Images: Getty Images, Alamy (Corythosaurus, egg)

There are so many problems researchers would have to overcome to clone a dinosaur

than just what its DNA says," she said. "A lot of the timing is dictated by genes and proteins that the mother produces during development. How is it going to get the developmental signals that it needs?"

Again, let's say that, somehow, the host mother was able to give birth to this creature. The resulting offspring would be a half-bird, half-dinosaur creation, Schweitzer said. But could this animal actually survive in today's climate?

"Its genes and proteins survived in a very different world," she said. "The carbon dioxide content in the atmosphere was different; the oxygen content was different; the temperatures were different — how is it going to function [in the modern environment]?"

Moreover, the creature's digestive enzymes might not work on modern animals and plants, and it wouldn't have Mesozoic microbes, which it likely would need to digest and absorb nutrients, Schweitzer said.

"[Dinosaurs] were designed to break down dinosaur proteins," she explained. "Or [ancient] plants, if you want to bring a plant eater back, which I'd highly recommend."

It would be cruel to bring back just one dinosaur for

" *They'd have to find a host organism to help clone the beast* "

our own amusement, she said. And it takes at least 5,000 animals to create a sustainable population with genetic diversity, Schweitzer said.

"How are you going to clone 5,000 T. rex?" she asked. "And, if you could, where are you going to put them?"

There are so many problems researchers would have to overcome to clone a dinosaur, Schweitzer said. "Getting the DNA, which we have not done — that would be the easy part," she said.

Still, she plans to continue her studies on dinosaur bone. And though cloning might be a pie-in-the-sky idea, she still thinks about it from time to time.

"To be honest, I'd really like to see a T. rex," Schweitzer said. "It would be very cool."

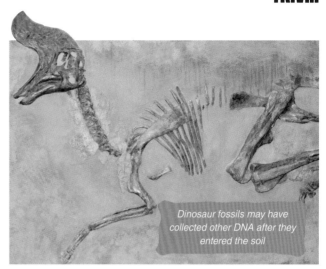

Dinosaur fossils may have collected other DNA after they entered the soil

Dinosaurs may not be able to survive by eating today's meat and vegetation

"Would it have been possible for us to live side by side with these huge beasts?"

COULD HUMANS AND DINOS HAVE COEXISTED?

What would the world be like if dinosaurs still lived among us?

What if the dinosaur-killing asteroid never slammed into Earth and the paleo-beasts weren't vanquished from our planet 66 million years ago? Would it have been possible for us to live side by side with these huge beasts?

"It's completely impossible," said Thomas Williamson, curator of paleontology at the New Mexico Museum of Natural History and Science, referring to dinosaurs ever being alive alongside humans — something that could never happen if the dinosaurs were to survive.

Though there were mammals during the dinosaur's reign of the Mesozoic era, these animals were small, no larger than the size of a house cat. It wasn't until the non-avian dinosaurs went extinct that mammals grew in size and specialty, eventually giving rise to the human lineage about 60 million years later.

"Dinosaurs had been around for over 150 million years when the asteroid hit, and were doing quite well up until that fateful day," said Steve Brusatte, a paleontologist at the University of Edinburgh. If the asteroid hadn't hit Earth, "I have no ▶

Artist rendition on the evolution of life from billions of years ago to today

MEGAZOSTRODON

The Megazostrodon lived over 200 million years ago and was one of the first mammals

doubt that they would have kept evolving and thriving."

If dinosaurs hadn't perished, "mammals would have never gotten their chance to evolve in that brave new world, free of their dinosaur overlords," Brusatte told us. "Without mammals getting their chance, then there would have been no primates, and then no humans."

Mammals originated about 220 million years ago, about the same time as the dinosaurs during the late Triassic. But dinosaurs got the upper hand — they diversified into thousands of species, spread around the world and grew to gargantuan sizes.

"Mammals stayed in the shadows," and none of them seemed to dominate their environment, Brusatte explained. Instead, early mammals mostly ate insects, maybe seeds and the occasional tiny dinosaur, according to fossil evidence.

Mammals didn't grow to the size of German Shepherds until after the dinosaurs went extinct

CITIPATI

Avian creatures were the only dinosaurs to survive after the asteroid crashed into Earth

Despite what you see in films and TV, humans didn't walk the Earth until long after the dinosaurs had died out

WARNING

DO NOT FEED THE DINOSAURS

" Mammals originated about 220 million years ago "

When the six-mile (ten-kilometer)wide asteroid collided with Earth, mammals and dinosaurs alike suffered great losses. All of the dinosaurs — except birds — bit the dust, and about 75 percent of all mammals died, said Gregory Wilson, an adjunct curator of vertebrate paleontology at the Burke Museum of Natural History and Culture in Seattle.

But there were some survivors.

"A few plucky mammals made it through the devastation of the extinction," Brusatte said. "These mammals seemed to be ones that were particularly small and had generalist diets, so they could survive by hiding and eating lots of different foods — traits that helped them endure the chaos after the asteroid hit."

Once the non-avian dinosaurs were gone, the mammals took over their ecological niches. Within a few hundred thousand years, mammals rapidly evolved (geologically speaking) into new species, diversified their diets and achieved new sizes. About 500,000 years after the dinosaur's demise, some mammals had reached the size of German Shepherds, Williamson said.

These spirited survivors are the reasons why there are more than 5,000 species of mammals today, Brusatte added.

"It's pretty obvious to me that none of this could have happened if the dinosaurs didn't die out," he said. "The mammals that lived with the dinosaurs had about 150 million years to make it happen, but they could never do it. But then, boom, right when the dinosaurs died, the mammals began to explosively diversify."

COULD DINOSAURS SWIM?

While they spent most of their time on land, there were things that motivated dinosaurs to enter water

YANGCHUANOSAURUS

> *They might not have been graceful, but they could swim nevertheless*

Some dinosaurs were amphibious but the majority spent most of their time on land

Whether a team of dinosaurs could win an Olympic relay race is up for debate. But they wouldn't be afraid to jump in the water.

All dinosaurs could swim, said Dave Gillette, curator of paleontology at the Museum of Northern Arizona in Flagstaff.

"They might not have been graceful, but they could swim nevertheless. Think of elephants or horses – they swim quite well even though their bodies do not look like the bodies of swimmers at all."

WHY SWIM?

Dinosaurs were motivated to swim by the same instincts that send a beaver or a duck to take a dip.

"They might swim to find food in water, to hide from predators, to cool off, to go from one bank or another, or even to swim across a river or a bay to a barrier island, and all the other reasons that an animal would decide to swim," Gillette said.

Like all reptiles, dinosaurs breathed air and had to take regular breaths, whether they were in or out of the water.

"Dinosaurs were surely just as adept at swimming, and just as talented at taking in sufficient air to continue breathing," Gillette said. "This all means that they had to be buoyant, too, so they could stay ▶

SPINOSAURUS

Baryonyx, like many other dinosaurs, are believed to have entered water to hunt prey

Images: Getty Images, Alamy (Baryonyx in water)

Mosasaurs like the Mosasaurus lived in shallow waters and would lay eggs on land

There were a number of reptiles living in the seas during the Mesozoic period

BARYONYX

close to the surface of the water, rather than sinking and drowning."

Although most dinosaurs spent a majority of their time roaming the land, some dinosaurs, such as Spinosaurus and Baryonyx, were likely amphibious. Both of these species were as large as T. rex and had an anatomy similar to that of crocodiles. They also had huge skeletal spines on their backbone that looked like a sail, but Gillette said those spines were covered with muscle and tendons and skin, and could not have functioned as an actual wind-catching sail.

Other than skeletons of swimmers, scientists have also discovered tracks of wading dinosaurs.

TRACKS OF SWIMMERS

"Some trackways indicate that dinosaurs 'poled' their way around in shallow water, like a boatman uses a pole to push a boat," explained Gillette. "Or, like the way humans push off and glide, then sink a little and then push off again, and glide..."

For example, in 2007 paleontologists from the University of Nantes in France came across S-shaped prints on the bottom of what was once a lake in the Cameros Basin in Spain. The unusual tracks suggest the animal's body was supported by water when it scratched the lakebed.

In 2005 in Wyoming, Debra Mickelson from the University of Colorado at Boulder discovered dinosaur tracks in what was an ancient sea floor. The footprints were left behind 165 million years ago by a dinosaur believed to be about the size of an ostrich.

"The swimming dinosaur had four limbs and it walked on its hind legs, which each had three toes," Mickelson said. "The tracks show how it became more buoyant as it waded into deeper water, the full footprints gradually become half-footprints and then only claw marks."

Dinosaurs weren't the only creatures showing off their swim strokes during the Mesozoic period. Many reptiles living during the same time as dinosaurs were restricted to living in the sea.

"Plesiosaurs, mosasaurs and sea turtles are all non-dinosaurian reptiles that lived in the sea in the Mesozoic and perhaps only came to land to lay eggs," Gillette said.

Spinosaurus were more adept at swimming through water than other dinosaurs

66 *Dinosaurs weren't the only creatures showing off their swimming strokes* 99

Images: Getty Images